THE DIGITAL CALL CENTER

Gateway to Technical Intimacy

PAUL ANDERSON
AND
ART ROSENBERG

Published by
Doyle Publishing Company, Inc.
5222 FM 1960 West, Suite 112
Houston, Texas 77069 USA
1-800-457-6459
www.doylepublishing.com

Publisher's Cataloging in Publication
Anderson, Paul, 1964-
The digital call center: gateway to technical intimacy / Paul Anderson and Art Rosenberg.-- 1st ed,
p. cm.

1. Call centers -- Computer network resources. 2. Internet (Computer network) 3. World Wide Web (Information retrieval system) 4. Internet telephony. 5. Customer services. I. Rosenberg, Arthur M., 1928- II. Title.

HE 8788.A64 1999 658.8'4 99-072185

ISBN 0-9653359-1-7

Cover design by Decoz
Edited by Kay Morton
Printed in the United States of America
First printing: 1999

About the Authors

PAUL ANDERSON is a Houston, Texas based writer, author and consultant who writes about and works with clients to develop strategies that take advantage of new communication technologies in this emerging age of access and information. In addition to consulting with industry leaders, Mr. Anderson has written and authored numerous reports, white papers and articles and has participated in many conferences on the technologies of customer relationships. Mr. Anderson is the author of the best selling book; "A Call From the 21st Century, The Technology of Customer Contact," currently going into a second printing. Paul can be reached at (713) 937-4125 or 73113.3352@compuserve.com.

ART ROSENBERG is from Santa Monica, CA and is a product/market planning expert, writer, and application consultant for call center and messaging technologies, having been involved in the development of the earliest attempts to automate call center operations in the late '70s with Delphi Communications Corp. Many of the "new" innovations being trumpeted by the industry were implemented for Delphi's call center services using their pioneering integrated multimedia system, including skills-based routing, screen "pops", multilevel desktop application displays, voice and text messaging, fax response, "blended" inbound/outbound call handling and others. Art consults and writes on a variety of customer technology issues and can be reached at (310) 395-2360 or artr@ix.netcom.com.

Both Art and Paul have contributed numerous articles on leading edge call center technologies and applications to such publications as *Business Communications Review, Voice Processing Magazine, Telephony, CommunicationAge, InterFace, TeleProfessional, Call Center Magazine* and others. Both have extensive portfolios consisting of white papers, research reports, and conference participation presentations for such technology information providers as Gartner Group/DataPro and Probe Research.

Table of Contents

THE DIGITAL CALL CENTER
Gateway to Technical Intimacy

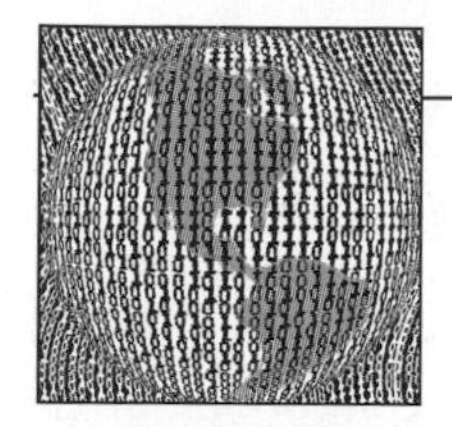

Chapter One
Embrace the Changing Edge

Introduction

We are taking part in the most extraordinary revolution in history. It is a technology revolution that has borne the age of multimedia; an age where the barriers of time and space no longer exist. It is an age where people from opposite sides of the world will soon be able to communicate face to face, just as if they lived next door to each other.

New networking capabilities, technologies that enable connectivity and bandwidth as well as the standards that drive new applications, are coming together in an environment of upwardly spiraling processing power and deepening technological penetration. Any number of converging technologies are contributing to this pace of change. There must be a dramatic fundamental change happening when toothpaste commercials now have world wide web addresses.

Technology is creating new interfaces, seams and edges on the peripheral, of which enterprises must adapt to or face the challenge of losing contact with their customers to the competitor that does. The internet is an excellent example and, while those sages and prophets of the internet have been fascinated by its property of having no edge, it does in fact have an edge. It is that boundary between those who are connected and therefore are valuable to the enterprise and those who are not connected and therefore are not relevant. The edge is a perilous place; it's the juncture where your enterprise network meets the great wide-open public network.

This fundamental change obviously has not escaped influencing the enterprise call center. Microsoft has a marketing message that says that information technology is increasingly forming the enterprises "digital nervous system." For the most part, the closest an enterprise can get to it's customers is the call center. This makes the call center the nerve endings of this "digital nervous system," the sensory inputs of the enterprise's "digital nervous system." As such, the call center is growing to become essential to the support of a customer's life cycle through the stages of acquisition, service and retention.

We should consider that the evolution of technology is primarily about the evolution of interfaces and that the call center, as the single most important locus of customer interaction and collaboration, is where we can witness the pragmatic application of the digital revolution. If it is not being used in the call center, it's probably not (yet) a significant technology.

The real struggle with enterprises today is not over information and the technology that creates it, but over relationships. When we design networks and customer interfaces for the enterprise, are we designing them to manage important information, or to manage important relationships? With the exception of those dependent on or responsible for the enterprise's customer relationships, it was not too long ago that information (data and the reports it generates) was more important than the relationships. But customer relationships too are vulnerable to the evolutionary influence of technology and we now see that technology executives are trending toward making their information systems more extroverted, with customer communication becoming the highest priority and customer satisfaction being the paramount justification.

Enterprises with extroverted-oriented information and communication systems are more successful and have greater productivity when compared to "introverted" companies. The extroverted company's productivity is growing comparatively faster as well.

Extroversion requires more accessible and open systems. Those executives that are evolving at the pace of technology know that they too must constantly work on integrating information systems and interfaces across the entire enterprise, dissolving previously impervious boundaries – those edges that have traditionally separated customer service from the rest of the business. The existence of this dichotomy is evident in the familiar gap between the data and the telecom disciplines.

Enterprise extroverts view the call center as a tool for increasing revenue by improving customer satisfaction; identifying and gaining new customers, enhancing the quality of the company's products and services and competing with speed by increasing enterprise responsiveness. In contrast, the old introverted model was one of customer service that put much more value on improving management's control of information flow, saving on labor (agents) resources and lowering manufacturing costs. Extroverts use call centers to improve their relationships with customers through the constant rapid redesign of business processes, while the introverts tend to be associated with little organizational change.

There has been a slew of studies, articles and reports in the past several years empirically demonstrating that US companies lose one-half of their customers every five years (known commonly in some industries as churn). The most successful strategy for reducing customer churn is to make it easier for customers to engage and communicate with your company.

Accumulating this experiential knowledge is driving competitive companies to extend customer engagements into extended customer relationships. The emerging view is that "extended customer relationships" now means the length of the customer's life cycle, expressed in the concept of a customer's "life-time value." Personalization, customerization, one-to-one marketing and relationship marketing are the buzz words of this emerging view.

Personalization is the integration of relevant customer databases to centralize all of an enterprise's content, policies and strategies to identify, generate and exploit new revenue streams from existing customers. The new power of technology increasingly allows enterprises to customize and personalize the experiences its customers have with the company.

The telephone has been the traditional interface of customer engagements for at least fifty years. Because of the deep institutional permanence of the telephone, many believe that it will be some time before the telephone will go away or even that its importance will diminish. This is not to say that the enterprise should ignore or discount the ever-growing importance of emerging interactive technologies. There is certainly an acute need, where there may not have been one in the past, to be increasingly aware of these emerging interactive technologies. Some will grow to be significant (for example, the internet) and some will fade into obscurity in spite of their hype (for example, Apple's Newton).

Knowing which technologies are potentially significant and those that are not is explicitly relevant to the business that you are in. As we evolve from the age of reactive customer service to the age of collaborative engagements, strategic decisions based on observations from the peripheral of the call center become increasingly more important.

From the Peripheral

To customer service executives, understanding the peripheral has had little importance since, for the most part, the telephone was the principle device used for customer engagements. The new age of access demands an understanding of, obviously, more than just the telephone (the internet really changed this one). Because interactive technologies are so temporal and fluid, but increasingly influential, one of the most valuable skills an executive can develop is the observation and understanding of the ever-widening peripheral influences of the call center. Management decision-makers rarely have the time to observe and examine all of the influencing technologies surrounding the periphery of the call center, but they no longer have the option of ignoring it either. Ignorance is a misunderstood word. It is not a synonym for stupid; it simply means lack of awareness.

The word "peripheral" itself is unique to information technology. There is no such thing as travel, food or shipping peripheral. It is from the peripheral that we find the basis for the way we do everything, from watching home entertainment, doing to our banking, handling our voice and printed communications or finding our way from city to city. Looking around the next curve, there are certain things happening on the horizons, the periphery, that will have profound impact on the way we do business.

We use the periphery to define what we are aware of without attending to expressly. For example, ordinarily when we drive, our attention is focused on the road ahead of us, the radio, a passenger, a car phone conversation, children in the back seat or the obnoxious slow driver in the left lane. Rarely do we concentrate on the noise of the engine. However, any unusual engine noise is immediately noticed, demonstrating that we are not so far from immediately noticing the noise in the periphery.

It should be made clear that "on the peripheral" is anything but "on the fringe," "insignificant," "inconsequential" or "unimportant". The danger with the evolutionary speed of technology's influence is in underestimating that what is on the periphery at one minute may be in the center the next.

Often customer service executives are so enchanted by the volume of information that an enterprise's information systems generate - and for most that's about all it generates so far – that they have neither the mind nor the time for observations of the peripheral. Peter Drucker has been warning executives for a long time about the dangers of being mesmerized by technology and the data it generates: "All great changes in business have come from outside the firm, not from within. But because saving time (creating a relationship) is an eternality, not capable of easily being measured in any of the typical productivity data, many experts and operational managers are missing the profound impact of technology."

Small jets are doing to the airline business what PCs did to mainframe computing, fractional horsepower machines did to turbines, mini-mills did to steel, cellular is doing to telephony, mutual funds are doing to centrally managed pension plans and what coming mini-generators will do to massive power plants. It is about power moving away from the machine-age center toward individuals of the microchip era (Forbes Magazine 6/97). In a couple of short years, the personal computer will be where a Silicon Graphics workstation is today. For example, the animated movie Toy Story was developed using 50 Silicon Graphics workstations. The same movie would have taken 47 years to render on a single Pentium chip today.

Increasingly, to understand what will be important next, one must pay attention to more fields of technology at the same time: chips, modems, PCs, software, fiber, satellite, internet, wireless and new applications such as electronic commerce. When it comes to call centers, it is the peripheral that is expanding; customers and callers are coming from all sides and faster than ever. Which of all of the peripherals should we follow if we are to know about the future? Some are predictable, some are wildcards. An example of an unanticipated wildcard with the new digital television (HDTV) is the problem, not with the technology (it looks great), but in the Sisyphan effort to build the broadcast towers. Quite literally, they could have seen the problem that there were not enough television towers and crews in the country to erect them, creating a huge backlog and most likely causing a failure of television stations to meet their mandated deadlines.

Even customers themselves loom as significant influences from the peripheral. Not so long ago, the customer call was a predictable metric defined by an Erlang value, but today it's no longer the call center that is increasingly digital, it is now the customer as well. The digitization of the customer is a penetrating factor influencing how businesses will perform and compete. How many call center executives are aware that in many US ghettos and housing projects, television penetration is greater than telephone penetration? If your customers live in ghettos, then you need to learn how to communicate with them through their televisions.

There are plenty of influential issues on the peripheral, some more influential than others. While an issue or technology may indeed be on the peripheral, until the number of people using a new technology reaches a critical mass, there are few real advantages to users (internet-based telephony, for example). Your telephone would be useless if you were the only one who owned one. The internet, television and radio would be useless without content; automobiles would be useless without paved roads and gas stations. But good content, highways and technologies don't occur until a critical mass of customers are ready to use them.

The use of language is another of the more underestimated peripheral influences. Language is certainly important to the call center because it is the most fundamental of tools that we use in call centers. There are two to three thousand languages spoken around the world, not including the strange language spoken by teenagers. Interestingly, almost one-quarter of all the English language has been created in the last 25 years, most of it having to do with information and technology. Ubiquitous words about the paradigm like interactive, virtual, models, feedback, infrastructure, and the internet fill, as Peggy Noonan is quoted later, "Our tribal language of jargon." The word "peripheral" itself is a relatively new word.

Even Microsoft Windows 95 itself now have its own dictionary definition. The definition of Windows 95: *noun.* 32-bit extensions and a graphical shell for a 16-bit patch to an 8-bit operating system originally coded for a 4-bit microprocessor, written by a 2-bit company that can't stand 1-bit of competition.

Language's influence on the business of call centers is significant and it tends to be inadequate to describe the importance of the langauge of the customer because executive management themselves are so saturated with jargon and acronyms, they simply can not hear. It is difficult to overlook the communications technology industry's particularly blatent penchant for developing new acronyms: TCP/IP, ACD/PBX, HTML, SMDR, SQL, LAN, WAN, SCSI, ATM, ADSL, ISDN, POTS and CTI to name a few. Using more elaborate and complex combinations of letters with catchy and zippy, "cadences of poetry is our industry's equivalent of a good old fashion pissing contest."

Peggy Noonan, a prominent White House speech writer, wrote in her recent book, "Simply Speaking," that, "The thing that has replaced accents in America is professional jargon, the shorthand language of shared knowledge. If you've ever seen ER, you've seen a good example of pure dialect, not of our tribe, but of our profession. Because our profession now is our tribe."

Observations from the peripheral demonstrate that even the Microsoft product PowerPoint is evolving from being a noun to being an adjective. Increasingly, it is used in reference to "powerpointing" observations, products or organization. When someone refers to "powerpointing" their comments, they do not necessarily mean to pull out their laptop and visually demonstrate their point, they are referring to using PowerPoint as a substitute to expand upon. One of the characterizations of "powerpointing" is a proclivity of call center technology vendors to promote the "powerpoint product," which, like the distinction between software and vaporware, often theatrically demonstrates products or services that do not yet exist.

History's Perspective

One of the most underestimated of the call center's peripherals is the one directly behind us - history. Along with the 35 year durability of Moore's Law, there also appears to be a certain amount of truth to the African proverb that says, "If you know your history, the future will trouble you."

1997 was the third decade of the PC, the 25th year of the microprocessor and the 50th year of the electronic computer. In 1975, there were only 300,000 computers on the planet. Today, we produce twice that many computers every week, or two every second. A 300 megahertz chip is worth more than its weight in gold. A program sold by Microsoft weighs nothing at all (Windows NT creates a profit of $90) and can be shipped instantly to any point on the globe. In contrast, the telephone is a 123 year old device that still dominates the interface landscape.

In the 1960's and 1970's, we ran businesses with big iron (mainframes) and batch processing, taking advantage of the capability to store and retrieve large amounts of data. In the 1980's PCs emerged, giving enterprise managers quick access to critical data and the ability to analyze, reorganize and process data much more quickly and effectively.

What history has to teach us in the call center business is that the staying power of the telephone can be attributed to the fact that we are, on a cultural whole, much more pragmatic than the popular media would have us believe. When the phone company first introduced touch tone calling, they charged a stiff premium; we bought it anyway. Now try taking somebody's touch-tone away from them. What would they have to pay you not to go back to rotary dial?

Answering machines were considered "insulting" by many in the 1970's, but by 1987 people were evenly split on their worthiness, this change took ten to fifteen years. Today, two-thirds of US households have them. It took eight years for credit cards to be embraced by consumers in the market place. In contrast, the internet has only been used as a sales medium for the past several years at most. First came email, and now the internet carries more messages than the postal service carries letters.

The first major step in increased data network access for the consuming public occurred over twenty years ago with the automatic teller machine and, for the first time, the general public came into direct contact with a computer-based application near and dear to their hearts, or more accurately their wallets. However, still today customers must travel to the ATM terminal and hope that the network is operating.

The transitory nature of the transistor is responsible for the fact that critical technologies that would have been meaningless 50 years ago (cellular services, fax machines, color printers, modems, internet service providers, music on hold, laser pointers, video conferencing), are now in the forefront of customer and enterprise engagements. Just as technologies evolve to the forefront, they dissolve into obscurity as readily. The VCR is a dying technology (being killed by DVD) as is the fax and paper checks. Just as assuredly, the cellular phone and the PC will die eventually.

One of the consistent lessons of technological innovation is that the emergence of standards stimulates both uptake and investment. In the early part of this century, convergence on voltage and current standards "sparked" an explosion in electrical devices and usage. Similarly, wireless communication standards created the rapid growth of cellular phone users in the early 90's. Standards today such as HTTP (Hyper Text Transfer Protocol), HTML (Hyper Text Markup Language) and IP (Internet Protocol) are driving the rapid growth of the internet.

The fact that the phone is over 120 years old testifies that the telecommunications industry's speed of change has been glacial. The fact is that it has only been relatively recently that our ability to manipulate data has exceeded our ability to communicate. Unable to no longer ignore pervasive competition, new legislation, new technologies and new customer demands, the telecommunications industry is having to rethink every assumption it ever made. One of the great historical ironies of the information age is that many of the telecommunications networks that the internet depends on are built upon industrial-age networks of railroad track and oil pipeline right of ways.

Knowledgified Customers

The most profound influence of technology on the enterprise has been the profound influence it has had on it's customers. Enterprise is increasingly finding themselves selling into an infomated world. Because of the fundamental availability of information, customers know more about you than they ever have before and they come to every transactions more "knowledged-ified."

The broad historical perspective establishes that power is now shifting from the enterprise to the customers and that technology, particularly the internet, is giving every customer the ability to survey the market place and make optimal purchases. Now enterprises are answering to empowered and smartened customers talking to equally skilled advisors.

What the internet does is enables is the availability to the customer of a vast supply of research and internal information on every imaginable item one would want to purchase. Retailing of every sort has gone from an information-scarce to an information-rich environment.

As interaction cost falls, goods and services providers will go directly to consumers or the consumers will do their own searching. In a single hour today, we can accomplish tasks that would have taken most of the day under the previous generation of interaction paradigms, such as buying a car or book. The savings that atom-based companies gain by bypassing traditional channels are more compelling than the enhanced services that characterize most on-line bit-based offerings. In the case of books (Amazon.com), PCs (Dell) and cars (AutoQuote), consumers directly benefit from the real costs that are taken out of the distribution system. Consumers today can rent cars, buy airline tickets and check in and out of hotels with little or no human contact. The internet and intranets have put meaningful technology power directly in the hands of employees, customers and business partners.

As the value customers derive from searching declines, intermediaries will have to enhance other aspects of their role, or create new roles to play. Some will build businesses around pure transaction channels; Charles Schwab is an example of this. Unlike the single call, or one-shot transaction, call centers are at the forefront of the emerging complexity of customer engagements (they are knowledgified now) and the demand for tiers of service (airline frequent flier programs) and preferential treatment for various classes of clients and customers.

In bit-based (information) businesses such as banking, finance and on-line entertainment, especially those supported by advertising, there is very little consumer cost to be reduced. The conclusion is that internet-based systems compete mostly on convenience and functionality. In bit-based businesses such as stock brokerages and travel services, the only thing that has been really replaced is the telephone call (ironically in fact it still being made via dial-up modem), essentially eliminating a customer service agent. The elimination of the middleman (an infomediary) creates lowered costs. Lower prices, not enhanced service, drive the demand. The customer's real electronic benefit can be found when buyers can save money.

Customer service-focused organizations are losing control of information technology because technology is no longer just a way to automate back-office processes or collect data; it has become a part of every enterprise's products and services, from the initial design stage to the customer service, support and retention. Customer interactions with the call center and all of the other types of self-help applications and mediums such as the internet, email and IVR make up the consumer's entire impression of the enterprise. Even though the amount of time saved and the ability to shop prices worldwide is a valuable resource, customer service is definitly lacking. Networking and telecommunications have matured and so have expectations. Smart shoppers are learning to identify and separate those that can service them – knowledgeably – or eventually these companies will perish.

It does not take much to realize that a satisfied customer is more likely to become a repeat customer, but until recently, technology that could significantly improve customer satisfaction was either unavailable or because of costs and complexity, limited to the very largest of formal call centers.

The new span and power of technology 'databases' makes it possible for every enterprise to remember each customer as readily as the customer can remember an enterprise. This is now easy to accomplish because the power of the processor brings the ability to remember each transaction and customer's preferences. Using multimedia server technology, these next generation systems can consolidate the connection, identification, management, routing and retrieval of virtually any type of customer contact, – voice, fax, email, or internet-based communications – into a single seamless process.

The key to management success in this age of access lies is spotting and analyzing key shifts in communication technology and then adapting and translating those changes into better ideas for helping your employees and customers improve core business functions. The challenge is understanding how communications technology is used to create stronger customer relationships - providing the best possible experience at the least expense to the enterprise.

Management

The economies of enterprise computing have, in the past five years, turned from struggling with scarcities of processing power, bandwidth and distribution to integrating with a networked society that is structured around flows of capital, information, technology, organizational interaction and relationships. Within a short period of time, many enterprises will face an embarrassment of having more information than they know what to do with and in the bloodthirsty business arena, it's either be in control of your information flow or get crushed under it. To avoid this, successful call center managers must increasingly become business managers, technical wizards, corporate team players, budget masters, long-term planners, educators and promoters. The new generation of call center managers must increasingly also deal with the pressures of budget cuts, mergers, downsizing, outsourcing and ever increasing workloads.

One of the most significant issues surrounding the enterprise's evolution into the new age of customer relationships does not lie with technology or the information access that it enables, it lies more often than not with the senior management. It is management style that is the most difficult to change in an enterprise because it involves altering ingrained, often institutionalized behaviors. Senior executives have no interest in changing what has made them successful and usually there is no one around to tell them to do it differently or that they must change. Those managers who anticipate and understand the fundamental nature of the changes ahead and actively reshape their business models in response will best be in place to exploit the opportunities. Those that do not will face a difficult transition from the legacies of the past to the intimate realities of the future. Anticipation is often said to be the basis for opportunity.

One of the impediments to management change is competitive uniphobia - a term coined by Watts Wacker in his book The Five Hundred Year Delta. Competitive uniphobia is fixation on competitive situations that, by their very character, are transitory in nature. Unlike the peripheral issues expounded on earlier, competitive paranoia is a useless exertion, with the exception of understanding how competitors themselves are using technologies to enable greater customer access. Looking at the competitive landscapes, particularly in the recently deregulated industries such as banking and utilities, an enterprise has no way of knowing who its competitors are, or will be, or what information about them it will need; it makes much more sense to focus on describing the competitive information the enterprise has today. Barnes and Noble never anticipated that an inventory-less company called Amazon.com would emerge as their chief competitive threat.

An implication of the technology revolution is the evolution of managers from infrastructure managers to information managers. This is a major cultural change for call center managers and customer service chain executives. In the past, managers managed the differences between data and information, now the paradigm is the difference between information and intelligence. Although many customer service managers try to relate the long term value customer relationships to revenue, asset deployment and operational costs, these are poor surrogates for the real implications of the "life-time value" of customers.

Describing and understanding any enterprise's current information environment can be a daunting and major undertaking. Describing who has what information, the various sources of information, how information and knowledge are used in work processes and an organization's intentions and objectives for this information is an essential task for competing. The rapid change and "growth" of an enterprise's information resources is becoming almost organic in its nature. In fact; new schools of thought express that information technology must be managed "ecologically" - growing, permutating and always in flux. By inference, the ecology of information becomes the ecology of relationships.

Information ecology is an abstract management metaphor, but a good one given the growing interconnectedness of the enterprise's information environments. As biological ecologies thrive on species diversity, so information ecologies thrive on information diversity. Still, true information integration won't happen without major changes in management approaches and organizational structure, particularly in the age of relationship management (a.k.a. one-to-one marketing). In fact, enterprises that wish to truly meet their customer's needs should not be directing people to any particular type of information, they should combine all of the information media available.

Another increasing characteristic of successful call center managers and customer service chain executives will be a trait (again framed by Watts Wacker) of intelligent disobedience. Intelligent disobedience is what seeing eye dogs are taught; essentially that they are to obey unless they have a better idea. Intelligent disobedience is not only a characteristic of the new generation of managers, but as an executive, the challenge will be in allowing a certain amount of intelligent disobedience in employees and agents. How many of us have had, from our perspective, exceptional customer service based on an agent or employee making a decision that probably was in infringement of a certain policy or procedure?

The opposite of embracing the search for intelligence in the mountains of information an enterprise can generate is avoiding it. Substituting activity for achievement, or discorporation, is a growing phenomenon where managers spend their days in "static observation mode" scrolling through screenfuls of data and management reports, bit by digital bit, searching for nuggets of revelvancies between call traffic, agent staffing, skills, trunks and trends.

Intelligent management, as opposed to informed management, distinguishes the difference between knowing how to effect a result and spending time achieving a result (substituting activity for achievement). As an example, there is a difference between knowing the formula to calculate how many golf balls it takes to fill an Olympic size swimming pool and exploiting other resources to do it for you (consultants, vendors, et al.). If calculating the number of golf balls in the pool is not your core competency, then you do not have to know that the diameter of the golf ball is 1.68 inches, that the radius is 0.84 inches, that a golf ball's volume is 2.48 cubic inches, or even that the densest packing of spheres possible is 74 percent. All you need is the answer of 68.28 million.

Rapid Prototyping

Complaining that technology changes fast is like complaining that rocks are hard; it's true but useless. It is also a little like expecting a bull not to attack you because you are a vegetarian. The future belongs to the organization that aggressively manages what they are not aware of, change. And just keeping up is not fast enough anymore. Today, an enterprise's ability to take advantage of the rapid hyper-evolution can be the most significant factor determining the enterprise's ability to keep up and compete with faster more nimble competitors. Take the idea of being open for business 24 x 7 x 365; not too long ago, this symbolized extreme competitiveness. Today, it is the norm for the blistering pace of a world open for business 24 hours per day.

Business is fast becoming a rapid feedback loop. Information sharing is what technology enables. Information from the feedback loop provides the ability to move quickly to offer new services and to unify whole enterprises, customers and vendors. What a free market does is allow people to make the cheap, fast mistakes necessary when developing complex systems and lifetime relationships. Information, unless it produces action, is overhead and one of the defining traits of a manager prepared to face the challenges of the 21st century call center will be the ability to adapt to the accelerating pace of change by discarding obsolete ideas.

Much of the current mindset about the organization of enterprises is inherited from an age of mass production and, in an interconnected world' small changes can quickly work their way through the entire system. Sampling and market tests, for example, are increasingly becoming limited, if not useless, because there is no way that all of the (new) variables that exist in the real world can be touched. In fact, one of the concepts that will drive the role of call centers in the future will be the idea of "pure response" marketing. Pure response marketing is the concept of near-instantaneous input of customer information, reactions and feedback into the enterprise's information systems.

Consider the speed of the stagecoach versus the train, versus a car, and now the space shuttle traveling at 36,000 mph. Jim Taylor, Gateway 2000's visionary, estimates that innovation in digital technology gets produced somewhere on average every three seconds. Product development approaches such as iterative prototyping and rapid application development are just as applicable to customer service initiatives as they are to developing widgets. Tom Peters makes the observations that the revolution of the last 25 years created the phenomena of Staples, Blockbuster, Home Depot, Circuit City and Wal-Mart, and it is his observation that these enterprises were born out of the competitive complacency of their traditional rivals. But, he astutely asks, "Do you think that the next 25 years are going to be more mellow than the last?"

Falling interaction and communication costs enable enterprises to execute strategies based on the concepts of mass customization. Not only will finding and reaching new customers become simpler, but tailoring products and services for them will be easier, faster and cheaper.

Look What the Net Has Dragged In

The internet is the technology that has emerged to be the one with the greatest earth-shrinking potential.

As we enter the first decade of the internet, we find that more than one trillion bytes of information (equal to 60 million typed pages) are posted on the internet each month. Internet traffic grows a hundred fold every 1,000 days. At the time of printing, the record for most web page hits in one week was from CNN Interactive, Atlanta, who said that its servers delivered 59.6 million individual page hits from August 30, 1997, to September 6, 1997 the week following the death of princess Diana. In one hour alone, CNN Interactive servers delivered one million page impressions. The number of internet users will grow by another 100 million in 1998.

According to a Booz, Allen & Hamilton study, a traditional bank transaction costs $1.07; the same transaction over the web costs about 1 cent. A traditional airline ticket transaction costs $8.00 to process, an electronic ticket cost of just $1.00. Federal Express loses money everytime they have to take a package tracking call. They actually make money when a customer uses the web-based tracking system. A personalized version of Time Magazine is available on-line to subscribers in Australia. Today, it only costs a few cents to create and deliver; in 1987 it would have cost $5.00 and taken four days.

Manual Castrell (author of The Information Age: Economy, Society and Culture) has defined the advance of the internet as "Rapture of the Net," "wherein this mystical network becomes a general purpose solution for social and cultural problems, a gateway to paradise. We're going to fix schools, reinvent government and link the world in an orgy of mass communication - and get rich in the process." The internet's growth will accelerate as computer prices fall and, although it is now dominated by American users, it will quickly acquire a more global reach. The span between nodes on the net is irrelevant since the speed of light essentially erases distance.

On the other pragmatic side, if you ignore the fact that after 125 years most of the 5 billion people on the planet still do not have telephones, adding on internet users could take decades. Some experts expect that the growth of the internet will in fact slow drastically and that by the year 2005 the internet's impact on the economy will prove to be no greater than the fax machine or Palm Pilot.

Solutions to the low bandwidth of ordinary telephone lines must emerge before the internet becomes the main medium for communications and contact. The battle is on between ISDN, ADSL or cable modems to supply high-speed net connectivity to the home.

Industry analysts and experts predict that web-based support will soon be the dominant means of customer service support. According to one of the leading internet analyst firms, Jupiter Communications, the internet will become a common method of customer support and they predict that by the year 2002, an estimated 50 percent of all on-line sites will use on-line support to augment their telephone customer service efforts.

As of March 1998, 15 percent of all call centers were web enabled. Some analysts estimate this number will double every year for the next several years. However, polls and surveys of Fortune 1000 enterprises indicate that most are not ready to integrate the internet into their call center operations. They do believe that the internet will grow to be a significant part of revenues and interactions in a few short years.

Strategic Danger Points

Technological progress is bringing about a massive increase in interactive capability. All modern forms of interactions – whether they be as simple as writing a letter or as complex as solving a customer problem in a call center – are being influenced by computing and communication technologies. The last several decades have brought remarkable innovations, but many feel that modern technology has failed to exploit opportunities to increase the quality and quantity of customer interactions and reduce their costs.

Information and data in the right hands at the right time can have a powerful effect. The same information misinterpreted or misapplied can be unuseful. How many useless web sites are languishing out there because someone got techno-duped into paying for them? There is an insidious side to technology and that is it often dupes us into changing our habits and overlooking common sense. Why use the internet for service if the ISP is always busy? The phone always works.

Investments in leading edge technology often reduce the very reliability and convenience that the "silent majority" of customers sorely desire and need. In fact, the majority of customers are often overlooked in the unimobileteledigicomlink babble of the more vocal, activist, functionality hungry customers. In satisfying this group, enterprises can pass up the opportunity to provide more of the reliability and convenience that mainstream users desire.

Too much technology can turn off customers. Systems that offer functions users don't really want and lack the qualities they find important, frustrate users. Not long ago, customers frequently complained that automation was putting businesses out of touch. Slow, unwieldy automated voice response systems were the biggest source of complaints. One of the risks from the rapid growth of technology is ignoring the significant proportion of the population, is in fact "down-teching." Add technology too rapidly and the enterprise runs the risk of frustrating a customer base used to personal service.

It is easy to have an appreciation of the potential of technology, but it is a problem when managers ignore what happens when technology misuses its potential. Take IVR as an example. In call centers the balance must be found that juggles the efficiencies of long menus versus the anger of customers stuck in them.

According to Forrester Research, Inc. in Cambridge, Mass., companies can trim 43 percent off their customer service and call center labor costs by offering some service on-line. In comparison, without internet-based solutions, the same cost will likely rise by 3 percent.

Using internet technology to reduce the costs per call or increase the number of calls handled reads well, but it is short-term thinking. The reality of this model is that it pushes the support and engagement burden back on the customer. The fact is that internet support does not necessarily reduce costs. Customers forced to exchange email with a product support person often takes weeks, when the issue could have been resolved in a fewer number of telephone calls. The individual transaction cost may be lower, but the frequency of contact is high.

Techno-duping (save a whopping 43 percent!) avoids focusing on customer satisfaction, not cheaper resolution. Competitive advantages no longer lie in saving a few dollars, but in fostering loyal customer relationships. If companies truly want to reduce their support costs, they should make products that need less support.

One of the pervasive, but often ignored issues in this age of access, is the underestimated craving of people for social interaction and contact in all of their engagements. One of the failures (or perhaps unanswered questions, since electronic commerce hasn't been around that long) of electronic commerce has been to faithfully recreate the social aspects of the traditional retail buying experiences. Barnes and Noble's competitive success against Amazon.com today lies in their offering of coffee shops and lounges, not in their book selection.

Understanding this characteristic of human nature, the striving to make technology more "intimate" is leading to some pretty humorous results. Researchers at Bell Labs have developed a Web page that turns yellow and crinkles in front of your eyes – just as a piece of loose leaf paper would after being handled by thousands of human hands. Algorithms automatically "age" the Web page based on the amount of traffic it endures. "Web pages are touched every day by thousands of people and there must be a way to convey the age of the page itself (using counters)," said Doree Seligmann, co-developer of the virtual paper. The Web desperately needs "signs of life and interaction in order to become more engaging." How do you intend to fill your customer's need for familiarity and intimacy? Or, are you avoiding it?

The Future

The trouble with the future, someone observed, is that there are so many of them.

The further ahead that communication technology futurists and experts project their minds, the deeper their insights are thought to be. Futurists would rather not predict what will happen next year. They prefer to think further, to a safer distance ahead, twenty, fifty or a millennium ahead. If they think far enough into the future, they are called, reverently, visionaries.

Futurists have a great line of work, they get paid to make broad generalizations, wild assumptions and baseless conclusions and all they have to do is be right about 20 percent of the time to be taken seriously. If only call center managers and customer service executives had that kind of luck or clout; If only 20% of your application attempts worked, you wouldn't be taken very seriously would you?

As we look to the 21st century, many of us try to identify trends that will have a major impact on every enterprise. The phenomenas that will change society as a whole – the aging of the baby boomers, new media like the internet and wireless phones, information delivery systems such as satellites – are surrounding us as influences. Rapidly falling communication costs will create equally rapid changes in the relationships between companies and their customers. As inexpensive communication becomes ubiquitous, geographic barriers between and among people, businesses and governments will diminish, opening up new commercial and social opportunities.

We have entered an era when an enterprise's anticipatory reactions are no longer governed by the domino effect, but by the "slinky theory" - a theory based on the premise that at any given moment society is either contracting toward consensus or expanding toward the exploration of endpoints. Unfortunately, " The shortest path between two points is always under construction." (Noelie Alito)

A whole new section of the global economy will see the communication capacity grow, as basic technologies such as PC and telephone service penetrate more deeply over the next decade. "A convergence of technologies is set to increase our capacity to interact by a factor of two and five in the near future. This enhanced interactive capacity will create new ways to configure businesses, organize companies, and serve customers, and have profound effects on the structure, strategy and competitive dynamics of industries." (McKinsey Quarterly)

Growing standardization will also encourage the integration of many technologies into the processes and methods of customer interactions. Just as the expansion in transcontinental railroad transportation compelled US railways to standardize track size in the early 1900's, the extra traffic created by lower interaction costs will force the emergence of new communication and information standards in the new millennium.

The general public will scarcely recognize that a revolution is underway because capabilities will come slowly and incrementally. The likely "revolutionary" technologies include ISDN, Asynchronous Transfer Mode and wireless. Local, long-distance, paging, cellular, entertainment and electronic commerce and internet access services which were at one time provided by different companies are now increasingly being provided by all types of telecommunications companies.

An example of the linear, pragmatic evolution of technology is movement of IVR toward ever more complicated and natural applications, of which the most natural is speech recognition. Speech driven autoattendant is certainly one of the killer applications of voice technologies. The next logical permutation after this is natural language recognition. Imagine what happens to the global opportunities when an enterprise can take calls from anywhere in the world and language is no longer a barrier to doing business.

In this new information age, information is currency and the discipline most likely to benefit from the emergence of the intimate transactional technologies is consultative support and service. "Infomediaries," which is exactly what agents in a certain context are, cater to the awareness that the time value, the financial value, and the relationship value of the next call is extremely important to the enterprise. This awareness alone, for management astute enough to embrace it, will drive the call center from the peripheral (there's that word again) to the center of corporate strategy; from cost centers to revenue centers, as the cliché goes.

Opportunities with intimate customer relationships and the call centers that serve them are sometimes difficult to anticipate, plan or foresee because many company's and enterprise structures today are still very much based more on assumptions about enterprise organization, than on the economics of interactions and communications (Egghead Software is an excellent example of this shift; closing nearly 50 stores to do business entirely on-line).

Anticipation is the basis of opportunity and, with each hard-won insight, the pragmatic manager will pause just long enough to plot a new course, designed to take advantage of what has just been learned. If you focus on too distant a goal, you may find an obstacle blocking the path that didn't appear on the map - better to follow the contours of the terrain. This is the critical function of having the enterprise enabled with technologies that make relationship and pure response marketing possible.

Conclusions

The effective use of technology is the price of admission to successfully compete in the next century. Managers and executives of customer-focused enterprises can either evolve by a thousand cuts, or experience revolution by big bang (fired). Adapting "out-of-the-box" thinking is absolutely necessary to become competitive in the new information era. Customer service decision makers must take risks when stepping outside of the usual paradigm of the "box" - because what is working today, won't necessarily work tomorrow.

The role of call centers will morph from simple large scale telephony exchanges to customer interaction centers surrounded by touch points of different devices. Enterprise call centers must adapt to a new stream of customer transactions that originate in cyberspace.

The tradeoffs that shape economic activity-enterprises trading off specialization against interaction costs, customers weighing current selections against further search costs, enterprises considering alternative configurations - will each find a new balance point.

The greatest evolutionary change for customer technology managers will come with the recognition that there is a fundamental difference between managing an information system and running an information ecology, just as there is a difference between operating a grape press and making wine.

Call centers are one of the few business disciplines that have delivered on the promise to improve productivity, information immediacy, raise the standards of customers service and meaningfully increase revenue. The future of communications is in direct and instantaneous contact with your customers and the public. It is certain that, with technology, we can boost social responsibility and consumer loyalty at the same time. Successful companies will drive toward customer enlightenment, not just enterprise enrichment; people, not profits.

"Manifest simplicity, embrace change, reduce network downtime."
The Tao of the Router
- computer ad

Chapter Two
The Decade, Demographics and Devices of the Technologies of Call Centers

Introduction

"It was not so long ago that people thought that semiconductors were part time orchestra leaders and microchips were very, very small snack foods."

– Geraldine Ferraro

The modern world's economy is in the early stages of profound changes. Two centuries ago, dramatic changes in the economy, precipitated by innovative developments in production and transportation, ignited the industrial revolution. Today, an upheaval of even greater proportions is about to be brought on by unequaled changes in the economics of communication and interaction - although it is not so much the changes, as it is the fusion that is occurring. The fusion (the call center industry's equivalent word is convergence) is happening between the power of computing and the bandwidth communication. What makes the digital revolution so perverse is its immensity.

The heat of this "fusion" demonstrates that we are no longer in the age of Aquarius. In fact, as Watts Wackers describes in his book, The Five Hundred Year Delta, this is the Age of Access. The Age of Access - a period we are already in, an era in which connectivity drives the access of everyone and everything and everyone to everyone. Access and connectivity are what call centers are about, not necessarily what they do.

There is an innovation in digital technology every three seconds. The result is that the price performance of modems is improving at 55 percent per year, fiber optics are proliferating at a pace of thousands of miles per day and television-quality bandwidth is being drawn out of plain old copper telephone wires. From mainframes to the desktop to networked computing of the 90's, incremental improvements in connectivity and bandwidth technology over the next ten years will exponentially multiply the interactive power of networks.

To avoid "paving the cow-path", the lure of "techno-duping" and the dichotomy of "good intentions, bad results", this chapter examines the underestimated influence on customer service and call centers by the advance of technology and the changing face of demographics. No longer the cliche "island in the enterprise", call centers' influences now come from the peripheral more than ever. Influences such as one-to-one marketing, the internet, Microsoft, core competency, 2¢ per minute long distance rates, wireless, PCs, customer demographics and many others.

Today, when we take a closer look at the misalignment between the technologies that are being used and the technologies being hyped, it becomes a critical, but often overlooked, practice to align business practices with new technology. What the executive strategists don't get, but that the operational managers responsible for the day to day operations of an enterprise's customer service function do, is a fundamental understanding of the economics and effects of changing demographics on who your customers are and, most importantly, how they will communicate with you over the coming years.

To baseline this perspective and bring some consistency to a market fraught with hype, if any particular technology is not found in a call center or is specifically being used to facilitate customer contact and access, chances are that it is not yet a commercially pragmatic technology.

Call Centers Are an Intellica

The biggest challenge of customer service chain executives today is maintaining *intellica* (en-'tel-Í-ka) a Latin term that is defined as the point between potentiality and actuality. This is what the call center has evolved to, the junction of old and new, past and present. Intellica is the point between smoke signals and internet telephony. Intellica is the point between rotary calls (of which there are still a substantial population in North America) and voice recognition.

On the potentiality side of the customer engagement is new technology; functionalities such as skills-based routing, IVR, the internet, IP telephony, workforce management, reporting and others. On the actuality side of the equation is the requirement for transactional efficiencies, speed of answer, quality of service levels, human resources, budgets, lines, ports and trunks. Intellica is the point between vendor push and customer pull. Your awareness of being your enterprise's "intellica" is an important and significant part of the evolution of your call center.

In spite of the blistering pace of change of a world open for business 24 hours a day, call centers seem to have the most pragmatic approach to the implementation of new technology, mostly because there is an embedded, intrinsic hesitation to experiment with customers. Managing for intellica is intimidating, and the unease it generates probably accounts for, more than any other factor, the perception of a slow adoption of technology in call centers. This "dis-ease" can't be measured, but we would agree that it is there.

Often, it seems as though industry has lost touch with the realities of running the enterprise. Call centers are in the unique position of having their feet in two worlds, the world of the pragmatic, functional and practical requirements of running the enterprise call center and the other foot in the future – the near future as well as the distant future.

Call centers used to be fairly isolated from the enterprise, an "island onto themselves." Now, as the intellica of your organization, the call center is the test and implementation bed for all sorts of progressive activities, strategies, applications and technologies. Some of these influences come from outside your company and you have little or no control over them; the internet and the growth of wireless are two examples. The redefinition of your enterprise core competency (although I will point out that your core competency, if you have not realized it already, is caller care) and one-to-one relationship marketing are examples of internal influencers.

As an intellica, the enterprise's call center is precisely where the rubber hits the road. If there were ever a front door or front counter to your company, the call center is now it. Call centers are, without question, at the heart of today's consumer-focused enterprise. The three or five minutes that a customer is engaged with your enterprise represents the "moment of truth"; the moment that if your customer's needs are not met expeditiously and professionally, your company stands to sacrifice quality in the customer relationship.

Those companies that have caught on to the "intellica" role of call centers realize that call centers are the closest that companies can get to their customers and the markets that they serve. The enterprise call center is the place where the practical implementation of the forward looking vision for today's customer engagements occur, because it already is the most critical point that an enterprise has to its customers.

The Uncertain Definition of "Call Center"

The term "call center" has devolved to meaninglessness and we have used it as jargon of convenience for so long now – more than twenty-five years – that it is just not worth trying to change, as technology in general has a deeply embedded collective agreement on this term. From time to time we see an attempt to change it into customer interaction center, or customer relation point, or contact center or various other terms. At this point, it's a term that's just too hard to change; some have tried using descriptors such as transaction center, information center, communication center, enterprise communications point, multi-media access center, casual or informal call center, customer care center and so on. Since everyone here today is in the business, we'll use it, but let's agree that it encompasses so much more now.

The term "call center" today simply does not fit what it now tries to define, which is the collective engagement of different media to effect technically intimate transactions with your customers. Today, the term "call center" conjures up a denotation of something formal and structured, high volume or high turnover, even though we would agree it rarely is any more. "Informal," "knowledge-based" and "casual call" centers are recent terms used to attempt to define these types of engagement "activities." It is consistent that the further from "formal call center" you get, the blurrier the definition of call center.

Call centers are now defined by what they do, not what they are. Do you have:

Contact Management?
Collaboration?
Customer Service?
Distributed "Net-workers"?

Any one of these enterprise functions uses CTI technology and applications developed from formal call center environments.

The Increasing Value of Time

Customers have become so used to the efficiencies created by technology that they are now impatient if a company's agent doesn't know everything about their histories and transactions as soon as the agent answers the phone. Consumers do not want to waste time not communicating meaningfully - you better have something meaningful to say.

Successful enterprises will adapt to the new paradigm of business drivers; the customers and their time. What technology is providing, be it through the telephone, internet or skills-based routing, is giving people more time to have contact with the people that they want to have contact with, while eliminating non-important, frictional encounters. The message is clear. Do not waste your customers' time by not communicating meaningfully; you had better have something to say and do it quickly, reliably and knowledgeably and consistently.

Time can be demonstrated as becoming a critical issue in the medical care industry. It is very expensive for patients to go to doctor visits or, even worse, visits to the emergency room. Research has shown that, of those patients who use a telephone to contact a call center and talk with the nurse advisor, 50 to 60 percent choose self-care with no visits. Another 20 to 25 percent opt for a next day regular visit with their primary provider.

Further, the patient/customer satisfaction levels are high - which is important in attracting and retaining more patients/customers. Following the same phenomena described earlier in this chapter, it is also less expensive to retain and cultivate existing medical provider members than it is to recruit new ones, by a factor of three or four to one. And, if the use of nurse "triage" programs can decrease office visits by 15 percent, then it is true that more members can be served with no addition of staff.

Time has become the most precious resource in the customer's life. We are entering a millennial generation where customers, like patients, are increasingly exploiting the efficiencies of new technology. Missed connections (ringing, then a busy signal), broken transfers and improperly handled customer requirements can put reliability expectations seriously at risk and can certainly cost the enterprise much more to repair than having done it right the first time.

For speed to work to the enterprise's advantage, it is critical that the call center be able to recognize a caller's identity or needs before the call is passed on to the agent. The efficiencies created by the "screen-pop" (customer data appearing on an agent's screen before the call is answered) can be dramatic. The American Automobile Association found that screen popping lowered the incoming call answer speeds from 45 seconds to 17 seconds. They also discovered that abandoned calls fell from 5,800 to 542 during a six month period.

Al's, "Moment of Truth"

What makes call centers important in the overall scheme of things is that they are a precise point where customer service chain executives have proven their power to impact company business. Call centers are often the first, only and last point of contact, and this is where building important customer bonds starts immediately or is sustained.

Total customer service means that when a customer calls, at that point of contact his problem, situation or inquiry can be resolved. This doesn't necessarily mean that it is being resolved by the same media or the same agent, but it means that the resources reside within that point of contact, whether it is a sales issue, a service issue, a warranty issue or problem billing; any of these can be resolved in that single point of contact and moment.

To compete, enterprises are going to have to use their call centers to solve more than just the customer's problems; in fact, solving problems is fast becoming just a commodity these days. Executives demanding that CTI be used to pump more transaction efficiencies are missing the point that when you have a customer for four minutes and you have his mind share, sure, access is important and cost reduction is important, but what is more important is the quality of the contact, not the quantity of the content.

Marketing costs are skyrocketing. A company typically spends approximately 15 percent of its revenues acquiring new customers and communicating with existing customers. It is increasingly difficult to communicate with customers for many reasons; one is increasing media fragmentation (150 cable channels).

The inbound call, if it is measured by some speed of transaction metric, is likely just a "dumb" call, not an invaluable or important call, but a simple problem-solving transaction. Customer service has traditionally been a reactive function with a goal of transactional efficiency. For the most part, the typical customer interaction has always been and is a simple transaction.

Companies pay huge amounts of money for a thirty second commercial; what they care about is what is going to happen in that thirty seconds. How come people in call centers are not focused on the same thing?

The subtleties between a successful television commercial and a failure can be very simple. Why are not the same resources that determine whether or not that ad is a winner do not get invested in like that spent measuring the impact of an ad?

As marketing changes become more intimate, marketing's role will be the weeding out of the customer you do not want, actually paying them to go away. Another change has been the development of marketing campaigns based on specific customer events. There is a great argument to be made for where advertising dollars should really be spent. It is not the ad, but the impression that has an impact; a good ad has a tremendous amount of impact versus a bad ad. A good customer service call has the same parity in customer expectation.

Overnight parcel delivery companies are a particularly noticeable industry where call centers are being put in as a secret weapon for the future. They realize that the phone call itself, in contrast to the television commercial, is a four minute ad. The interaction event between the customer and the enterprise is not a transaction, it is a four-minute commercial that will leave an impression not less significant than a 30 second ad on the TV. These companies have learned to use this traditional transaction event as more powerful and impactful than any television ad.

The package delivery companies take tens of millions of telephone calls every year, each lasting a couple of minutes. Call centers are highly articulated environments capable of measuring the effects of subtle changes in people, application and technology. In fact, call centers are experts at measuring efficiencies, but how should a company measure the mind share of each one of those calls - each lasting several minutes or more? Particularly in comparison to the mind-share gained advertising on CNN, which on average has only 350,000 households watching at any one time.

As North America and the rest of the planet rapidly shifts to a digital storefront, it is the customer's service expectations that will have the greatest consequences on the call center.

Creating Relationship Expectations

"Mass advertising and marketing is an industry of alchemists who routinely use the sorcery of slogans and images to turn toothpaste into sexual lure, athletic shoes into totems of hipness, and gasoline into expressions of manliness. Yet in spite of its ability to transmute the base metal of human insecurity into gold on behalf of others, the advertising industry has been saddled with a decidedly poor image of its own and advertising executives were narrowly defeated for last place in terms of honesty and ethics by members of congress and used-car salesmen." (Unattributable.)

The point is that we just do not buy it any more.

In 1996, more than $300 billion was spent on consumer advertising and promotion. In 1998, the average consumer will see or hear 1 million marketing messages - this is almost three thousand per day. No human being can pay attention to this every day. It is an assault and we are turning it off in favor of things we are particularly interested in. "Winston tastes good . . . like a cigarette should." This commercial was aired thirty years ago. If you wanted to build a slogan like this, you would have to burn the message into people's brains and it would cost a huge amount of money. Do you know how much Nike spends on Just Doing It? Be like Mike, Obey Your Thirst. The most powerful force in the universe is denial.

The biggest problem with mass-market advertising, like Super Bowl television commercials, is that it fights for people's attention by interrupting them. Telemarketers interrupt dinner and a print ad interrupts the article you are reading; bingo cards fall out of the magazines you are already subscribed to. A thirty second spot interrupts "Seinfeld" and spots interrupt sports.

These interruptions steal our time. In fact, according to people like George Gilder, time is becoming our most precious commodity. We don't think of it, but one of the largest and most successful call centers of all time is Domino's Pizza. Domino's Pizza is entirely based, not on their pizzas, but on their ability to route and network phone calls. There are many other better pizzas than Domino's; there just are not many that are faster.

Disney is not only a global leader in theme parks, they are experts in customer service, as they have perfected the superb design techniques of persuading customers to wait in line, which has an interesting parallel to the call center experience. Keeping customer needs at the forefront of the contact experience is what Disney has learned. For example, telling those waiting in line what their expected wait time is at numerous points. They keep attractions available to distract customers from focusing on long lines and they also move the line back and forth to give the impression of movement and progress.

Customers don't buy into this stuff as readily any more. In fact, the average person today is tired of being assaulted and is noticeably beginning to demand that every engagement be speedy, accurate and consistent. The "bar" of expectations is no longer set by advertising slogans, but through the real experience of the emerging complexity of customer engagements (the internet has guaranteed that customers are more "knowledge-ified" than ever) and their increasing expectation for tiers of service - an expectation developed by airline frequent flier programs, for example.

Communication is more relevant to everybody when is it is initiated by the customer as opposed to when the enterprise decides to initiate it.

Call Centers and Customer Satisfaction

All customers are part of the shift from transactional efficiencies to technical intimacies. This is not the same as the shift to supreme technical proficiency. It is the quality of the contact, not the quantity of the content, that matters. There is a subtle, but sublime difference. We have failed miserably in understanding that customers have no way of knowing our expertise; but they do know the quality of the contact and that is what they increasingly compare our enterprise to. Call centers consistently confuse quantity of content with quality of contact. We now know empirically that the greater the quality of contact, the greater the retention. Mail order firms are providing better customer service than their retail competitors who face customers directly. Their agents (clerks) are better trained and friendlier.

Technical support operations often have a chain-like property of being no stronger than their weakest link. What differentiates support from service is the higher knowledge component required for support. These are predictable routine engagements. Support is more complicated; the customer is looking for answers, not information. When this link is broken, the customer's perception is that all aspects of the enterprise's customer support and service are broken. After getting nothing but busy signals on main support lines, attempts to access alternative support channels were met with long delays, inappropriate canned, pat answers or no response at all. Busy toll-free lines are expected and not surprising; what is surprising is the difficulty users have in finding solutions via other methods.

Competitive companies are extending improved customer engagements into life long customer relationships because it does not take much to realize that a satisfied customer is more likely to become a repeat customer. They know intuitively that small movements in customer retention can equate to large shifts in revenues. In fact, a small change in only a few data percentage points in customer retention can equal millions of dollars to the enterprise.

In a vertical market segment heavily dependant on call centers, direct marketers in particular have a myopic tendency to focus on prospecting for new customers at the expense of maintaining ties and maximizing relationships with existing customers. Besides, the downstream effect of a dissatisfied customer can be worse today than ever. Marketing's role evolves into intercepting the customer complaints before your competitors do.

Understanding your customers is the key to loyalty and retention. The importance and cost effectiveness of keeping the customers you already have is well documented. The average US corporation loses half of its customers each year. The cost of regaining a lost customer has been calculated at 12 times the cost of acquiring a new one, and finding a new one is five times as expensive as keeping an existing customer. It takes three to five satisfied customers to reach the same number of people as one unhappy customer. More than 90 percent of dissatisfied customers will not come back, regardless of the effort you make.

If customer satisfaction is ranked on a scale of one to five, from completely dissatisfied to completely satisfied, the fours, though satisfied, were six times as likely to defect from a company as the fives were. This was demonstrated in a study conducted by the Xerox Corporation. This contradicts what we are used to believing about customer satisfaction and loyalty being tightly related and linear.

Customer loyalty, we now know, rises little as satisfaction rises and customer loyalty only rises sharply at the highest levels of satisfaction (the fives). "If service is the key to customer loyalty and customer loyalty is the key to sales and profitability, then a crucial question for business is finding the best strategy for maintaining loyalty." Nadji Teharani, TCCS, 2/98, p. 24.

These "fives" are also called the Pareto customers, after the Italian economist who discovered the 80/20 rule (the rule that 20 percent of your customers account for 80 percent of your revenue or profits). If your Pareto customers are only two percent of your customer base, but they are responsible for 50 percent of your business, then the corollary applies that you can affect 50 percent of your business by working for those two percent.

Technology critical to satisfying customers has only relatively recently allowed the enterprise to track and analyze customer information. In call centers, the center of gravity is shifting from measuring the efficiencies of calls to the satisfying of customers. Customer expectations of consistency, effectiveness and efficiency of the customer interaction with the enterprise is increasingly being measured, not by the number of requests answered, but by the quality of the interactions. As a result, the enterprise's performance is being measured against more strategic measurements, such as customer retention, customer satisfaction, consistency in service level, application of best business practices and ultimately, revenue growth.

Many call centers have long been motivated by the reduction of costs, which leads to call avoidance. Inside the call center with all of the rigorous personnel and system performance measuring metrics and applications, it is much easier to measure the impact of saving a second or two from a call. Many call center managers still value transactional efficiencies over customer intimacies and they have not thought of using each "transaction" to pry a little deeper into what the customer may really believe about your products or service. This is an opportunity to query the customer about what they are thinking about your competition. Many enterprises would be surprised to discover who their real competitors are. Focusing on solving the customer's problems can still be accomplished on the call, but while they are on the phone why not do some other things?

The Technology and Data of Contact

". . . keeping up with the pace the blistering pace of a world open for business 24 hours a day."

It takes an awful lot of infrastructure to support the blistering pace of a world open for business 24 hours per day. Few people realize that the largest deposit of copper in the world used to be in Chile; in the last fifteen years it has shifted to New York City. There is now 43 million tons of it in North America, although we are lightening up to the rate of several thousand miles of fiber optics per day.

The universe may be expanding, but thanks to the microchip, our planet is definitely shrinking. Before we get into some detail, I want to point out that all of the marvelous technology affecting the technology-based delivery of customer service is brought to you by the increasing power of the incredible shrinking microchip.

I am certain that you have seen this in other presentations, but I still find that this figure keeps us in perspective. We are using big numbers today, numbers bigger than we have ever used in our entire history. We use numbers we never used before in just my lifetime: millions, billions and trillions. To give you an idea of the magnitude of these numbers: one million seconds is 11 days, one billion seconds is 31 years and one trillion seconds is 29,000 years before the birth of Christ.

Intel processors have gone from a thousand transistors (and one-quarter mile of wire) to a billion transistors and seven miles of wire! Each Pentium has three million transistors that can be operated in more different ways than there are particles in the universe. There are no fewer than 15 flavors of Pentium computer chip and this doesn't include the various flavors of Cyrix.

There is a direct, although distant, correlation between the rising expectations of service from customers and the dramatic falling costs of interaction and communication. Today, a single raw Pentium MMX chip is only $316.00! (Although the factories to make them cost more than a nuclear power plant to build.)

The performance of Moore's Law (processors double in power every eighteen months) has now held true for 35 years. For $200 today, you can buy 350 times as much PC hard disk as you could for the same price ten years ago.

On the power side of the curve, these same cheap chips are exponentially increasing in power. Intel scientists have recently built (at a cost of $55 million) a computer so powerful (9,200 Pentium Pro chips) that it has passed the tera flop barrier – more than one trillion calculations per second. This is two and one-half times faster than the nearest supercomputer. In a single second, this computer calculates as many calculations as the entire US population (220 million people) could do on hand calculators over a time span of 125 years! MIT recently unveiled a 26 giga flop machine and IBM has come out with a 4,099 processor supercomputer that can handle three trillion tera flops. IBM's new supercomputer runs 15,000 times faster and has 80,000 times more memory than an average desktop PC.

The change in the disk drive market is also equally astonishing. In the disk drive industry, the size of drives has consistently shrunk from 14 inches to 8, 5.25, 3.5, 2.5 and now 1.8 inches. In 1990, the common PC disk drive that could hold only ten thousand double spaced typewritten pages now can hold 1,450 average sized novels, or more than 725,000 double spaced typewritten pages (an 18 story stack) on a single square inch of disk! These disks will be commercially available in two to three years. IBM expects to reach the ten billion bit level by the end of the decade. This is important to the technologies of customer service because all of this disk space will be needed to store all of the customer information we will be collecting.

Not only are we able to process customer data faster and store more of it, but we can send it faster as well. Over the past decade the rate at which data can be transmitted over a telephone line has increased four fold. It is predicted that over the next decade it will increase 45 fold. The time it takes to transmit 100 kilobits of data fell by 75 percent between 1985 and 1995; by 2005 it will have fallen another 97 percent. In a span of six to eight years, the price per kilobit today of 0.31 cents per kilobit will plummet to 0.004 cents per kilobit by the year 2005 - a fall of 98 percent. (McKinsey Quarterly, 1997, Number 1, p. 11.)

Computing power will not be a bottleneck. Cellular, satellite, cable, and wireless technologies will expand the bandwidth available for communications. Processing power, memory, hard drive access and modem speeds will get faster, cheaper and more abundant.

These expectations of service are a direct result of the power of the computer; we now can keep track of every transaction, conversation and detail about our customers. Transactional data from sales and service "touch points" across an enterprise create an (often over-) abundance of information about the customer. This transactional and dialog detail enables an enterprise to execute strategies based on the concepts of mass customization or relationship marketing.

Based on the power of processing, a true knowledge-sharing company can now store, mine, and exchange any form of data including images, video, audio, graphics, animation and text. It can also access and exchange information over a variety of networks such as the internet or corporate or campus networks. Soon we will all have access to the ability to manipulate any kind of data you can imagine. Transactions, text, numbers, multimedia, web pages, geo-spatial maps, time series data and even data types that you create.

The Revolution of Contact

Technology's pervasive influence affects call centers in many ways, not just in the enabling of the media customers use, but in the demographics and geographics of who they are as well. Technology has defined the evolution of American cities and where we live.

In 1997, for the first time, more people moved out of cities than moved in. This is causing an instability we haven't seen since the change from the agrarian society to the industrial revolution because technology has reversed the need for centralization. Our customers are becoming more dispersed. This is an important relevance to the anticipation of the interfaces that they will use to engage the enterprise for service. When we consider a world where our customers are connected to the enterprise through vast fiber optic-based networks, the observation of customer demographics (where they live), says that the future won't be based on fiber-to-the curb, but fiber-to-the-farm. This is not likely to happen anytime soon.

Cities have always been shaped by the state-of-the-art transportation of the time. Donkeys, horses, wagons and Roman war chariots created Bethlehem; the automobile built Los Angeles; jets built Denver and Houston. PCs, TVs and the internet have become today's state-of-the-art transportation. We now move information without supporting the need for travel. The whole automobile industry was invigorated by the development of the interstate highway system. Alongside, literally, of the interstate system came an entire service station industry to service those autos. Remarkable parallels exist between the development of the interstate highway system and the emergence of the information super highway.

Both the PC industries and the automobile industries were started by hobbyists, pursued by fanatics and then embraced by the broad population. Sometime this year, IT will surpass the auto industry's share as a percentage of the US gross domestic product. In fact, Ford makes more profit from its information and data-intensive finance and leasing units, not from selling cars.

Techno-duped and Paving the Cow Path

When we look at the rush to incorporate new devices, media and the internet into the enterprise, we tend to underestimate the pragmatic nature of our culture and society in general. Trains, VCRs and computer keyboards are excellent examples of the persistence of legacy technology. You would think that after 120 years a telephone would have more than 26 keys to represent the alphabet. We, as a whole, do not change our interfaces quite so easily. In fact, accounting for the growth of the wireless phone, the telephone is still the fastest growing device in history - nearly 35,000 new subscribers per day are being signed on in North America alone.

The world is out to solve transactional problems with computers and they are not facing as much progress as they expected. It is often hard for people to switch to a new technology that is only slightly better than the old. A phenomena that can be described as path dependence has since become the basis for the tendency of high tech markets to encourage monopolies that may not necessarily offer the best technology (Microsoft, for example). The idea is called path dependence after the notion that once you are started down a certain path, it is hard to get off. Modern railroad gauges are an example. They have the archaic characteristic of being the same width as two Roman war chariot horse's butts. Often cited too, is the VHS recorder, where the VHS won out over the superior Beta format because of early market maneuvering, not better technology (all the good movies and all of the early adult content films came out on VHS). We don't even let go of our toys so easily either. The "Slinky" toy was invented in 1945 and by its 50th anniversary in 1995, there had been more than 250 million sold worldwide.

The telephone persists because of an institutionally embedded "use-to-it-ness." One of the criticisms, as well as strengths, of the traditional call center technology vendors is that they are influenced by executives who buy into the notion that free markets choose the best technologies. According to long held Ivy League economic theory, in free markets it should be impossible to have the persistence of inferior technologies.

Historical events have a large and unacknowledged influence in shaping economic choices, and we find ourselves today getting locked into inferior technologies and products because of distant and long forgotten events. The computer keyboard is the most used tool in the call center, perhaps as much as the telephone itself.

But let's look at the QWERTY keyboard, named for the first six letters on the top row of keys. The keyboard was designed in the 19th century to deliberately slow down typists because the earliest manual typewriters tended to jam at higher typing speeds. QWERTY solved this by placing frequently used letter pairs far apart. In the 1930's, a university professor named August Dvorak developed an alternative keyboard that was much easier to learn and type on, but nobody switched to it because the QWERTY keyboard had such a head start. In fact, estimates were that a company could recoup the cost of retraining a typist in as little as ten days but this retraining never took place because, in effect, the market had already settled on the "wrong" system.

The sensitivity here, and we recognize it, is the risk of losing the "human touch." For the past twenty years, delivering customer service over the phone has remained largely unchanged. While voice response technology has improved call centers by reducing live agent calls, its use has been limited to relatively simple transactions. The longer and more complex transactions influenced by ERM (Enterprise Relationship Management) and web knowledge-afied customers continue to require live agent assistance and the demand for agents (although different kinds of agents) will continue to increase.

In the last five years, there has been a lot of discussion and hype around CTI and many think that we are missing the boat on what the real focus is here. The focus should be on the actual interaction with the customer, for the three to five minutes that the customer is engaged in a conversation with the company; that is really the only time that the customer cares about. The enterprise is either going to make or break the event, whether that is selling them something or making them happy, in that three to five minute window.

The challenge with technology is that too much of it can turn customers off. Systems that offer functions that users don't really want and lack qualities they find important, can frustrate users. What the "advanced intelligent network" has delivered is CTI, telephone tag, voicemail jail, dropped connections and telemarketing invasions.

CTI is a small component of an overall solution. CTI gets more press than implementation and is a concept that is grossly misunderstood. CTI does provide some level of productivity, but the world needs to focus on the "moment of truth" and this is when the customer is engaged in the conversation with the enterprise. This isn't just for the phone conversation but also exists in other media as well, like web site or email.

Computer telephony applications will be to the 90's what LANs were to the 80's: high margins, great customer-pleasing solutions and customers happy to pay those high prices. For computer and network resellers, they are going to have to learn telephony, i.e., get trucks.

There is generally a lack of focus on the right issue with the call center. Harry Newton has done a great job of taking a component of a call center that happens to be sexy and making it a major event. So much so that this has emerged as the center point. Vendors are developing and delivering transaction-based systems because of the "cultural" belief that computers are for transactions; they did accounting for years and they should be used for transactions now and this is what most CTI data guys sell.

Take IVR as an example, where there are issues about the efficiencies of long menus versus the anger of customers stuck in them (or even worse, voice mail jail), generating a suspicion of evasive antagonism. The expectation of IVR to shorten calls on the front end has not proven to be true; IVR actually cuts time off of the back end of the call in reducing the number of calls handled by each agent.

We are starting to see all the way up to the enterprise's "C-level," an increasing demand to be able to bring telephony into the mainstream of business application development, treating the telephone as just another IO device in their business applications. This is partly influenced by the challenge that the internet has posed to companies. Information taken from already-written verbatim hard-copy documents and presented the same way to each customer.

The global emphasis on electronic commerce and the use of the internet as a delivery vehicle have sparked the development of new CTI applications that offer tremendous opportunities to call centers. Unfortunately, the sad fact of internet advertising is that most companies building a web site use "brochure-ware." Many corporate web sites have evolved from electronic versions of glossy brochures offering little more than a profile of the company to an emulation of the call center that offers an alternative delivery channel from which transactions can be conducted in a self service mode - without having to speak to a live agent. Now, increasingly, we are also seeing that in that same application, they want to be able to integrate the live voice of the call center agent, the recorded voice from a voice response unit and interaction from the web or from email in a single cohesive customer environment.

The Data Guys and the Telephony Guys

A vast array of forces - including pervasive competition, new legislation, new technologies and new customer demands - are causing the telecommunications industry to rethink every assumption it ever made, from mainframe central offices, to PBXs and vertical integration by telecom manufacturers; from closed systems to open systems.

There are several significant differences between the two industries, computing and telephony: telephony is government regulated, computing isn't. In telecom, marketing is an endeavor in asking the regulatory authorities for a rate hike without pissing off the public so that they notice. In computing, marketing is building a better mousetrap. The dichotomy is evident by observing that IBM lost a tremendous amount of money on telephones and AT&T lost a lot on computers.

Both the information (data) and telecommunications (voice) revolutions are converging toward the same objectives - a single network infrastructure that will carry voice, video and data traffic; but both have different visions of how it will be accomplished and how long it will take. Those from the voice world believe that melding the voice capabilities into the data network will take years; some even suggest that it will take decades. Many executives on the data side expect that the annexation of telephony into their networks could take as little as just a few years.

The telephony guys believe that the data world has drastically underestimated the effort and the expense of bringing latency-sensitive voice traffic to data networks based on the Internet protocol (IP), while at the same time maintaining the expected reliability and quality of the telecom network. “They don’t even know what they don’t know yet,” said one telecom executive referring to the inexperience of the data world in the telecom world.

Where does this difference in beliefs come from? Their cultures. Time just moves a little slower in the quality over quantity voice world. New service rollouts in the telecom world come out at the speed of a glacier. It is not uncommon to find telecom switching gear, ACDs and PBXs that have been in use for ten to fifteen years. This mind set is the opposite of the data world side of the equation, where things move so fast that nothing is ever settled and investments obsolete themselves before implementations. The role of the router for example, has unarguably changed several times in the past several years. Traditional telephony devices such as PBXs, IVRs and ACDs are already being absorbed into the data networks.

There is no doubt that the data guys are pushing ahead faster into voice than vice versa. The data world appears to be rushing to see that the merger of the voice and data networks comes together before the telephony guys can complete their crash course on packet-based IP networking. Meanwhile, the telecommunications world is looking for the time to build up their data arsenal.

Largely left unsaid and un-media-cized, is the fact that the data world is dealing with substantial legacy issues of their own. In addition to their own legacy issues, there is still quite a bit of foot dragging when it comes to expectations of reliability and quality. In fact, based on the issues of reliability and quality alone, the strongest suit of the telecommunications industry is their deeply embedded techno-cultural delivery of reliability. Don’t discount the voice world, any down time is an emergency. Don’t write off the telecommunications industry as being the source for many of the innovations, applications and technologies that will provide the solutions of customer relationships in the information age.

Contrary to the perceptions generated by the popular media, the convergence of the data and the telecommunications worlds will see the significant, mostly unspoken, influence of telephony guys with laptops. Don't throw away those PBXs; they are the most reliable systems we have.

The Tensile Strength of Institutionalized Reliability

Because a commercial program of the complexity of the software used to run the space shuttle would have 5,000 errors in it compared to the 11 bugs found in the last 11 versions of the shuttle's programming code. But 99.999 percent isn't even good enough for the space shuttle. This software controls the space shuttle, a $4 billion (there's that billion again), the lives of a half dozen astronauts and the dreams of a nation. Even the smallest error in space can have enormous consequences; a bug of only two-thirds of a second while traveling at 17,500 mph puts the shuttle three miles off course in under thirty seconds.

The Denver airport couldn't open because of software, and the B2 bomber couldn't fly on its maiden flight because of software bugs. How many of you here don't think you have bugs in your computers? When the office LAN or voice mail goes down, we are inconvenienced, but we expect it to happen occasionally. No one expects his telephone to go out of service three or four times a year. We are not conditioned for web brownouts; AOL is in a lot of trouble right now over this.

Many IT professionals are saying that telecommunications are within the ken of IT; no way is down time an emergency.

Southwestern Bell is so proud of their reliability expectation that they even put an ad in the newspaper last month that reads: "Southwestern Bell Engineering has produced a network that's 99.999 percent reliable." When you don't get that dial tone, people notice. What kind of equipment do you think SW Bell uses to achieve this 99.999 percent level of confidence? By the way, SWB uses Lucent and Nortel-built switches in their central offices.

Unlike the space shuttle's software, it is a little harder to reboot a customer call. You want this kind of reliability in your call center, don't you? Networking and communications have matured, and so have our expectations. We live in a culture that has been conditioned to the reliability of the software used for dialing 911 for emergency assistance. We certainly don't think of dialing the Poison Control Center's emergency help on line forum on CompuServe or the local fire department's web page, if our child is choking. When we pick up the phone, we expect a dial tone not 99, but 100 percent of the time. Does your call center or company web page have that quality of service?

In computing, reliability is 23 hours a day; telecom reliability is less than six seconds of down time per year. You don't turn phones off for backups and upgrades.

Mixed Media, Mixed Results

Today's approach is piecemeal - setting up distinct operations to handle telephone inquiries and point-of-sale that are separate from web email teams, fax and other product and response groups within the enterprise. This approach makes for very inconsistent responses, fragments potential customer relationships and precludes the use of customer data across the enterprise units. Customers easily cross over between email and live contact without starting over, not to mention multiple business units within the enterprise. And the customer's account information shown on the internet may not, and frequently does not, match up to the VRU, agent's interface, branch office or point-of-sale.

The challenge of these technologies is to avoid serving customers inconsistently due to multiple highly independent departments and technologies in the enterprise. Customers want a consistent interface with an enterprise, regardless of the entry point.

While the web and the internet are similar to email, each has its own dimension with unique issues to consider. The web mail, "email us" button, is completely integrated as a part of the enterprise customer interaction system. That the internet-enabled interaction point is not established as separate customer stovepipes. And making it user friendly and creating browser echoing for agents so that the caller and the agent can be looking at the same information simultaneously.

Yes, there remain challenges in weaning current non-internet users off of traditional contact mediums such as the telephone and the fax and onto the exponentially diverse internet and email.

The internet can be used to strengthen current relationships and make them more efficient by lowering cycle times, order management costs and inventory levels by increasing the quantity and velocity of the information shared. There are drivers for the internet - other industries that have nothing to do with technology are being built on IP. Banking, gambling, travel planning, games, pornography and publishing are examples of the few successful applications of technology.

Virtually every computer is networked today and is a common mechanism for delivering information to people, whether the information is generated internally or outside the enterprise. On the face of it, electronic commerce's ability to reduce transmission and transaction costs makes it a natural for hard times. For the internet to work in this strategy, it must be scalable, be linked to business needs and deliver a real, sustainable advantage. The best markets are the most technically sophisticated. Putting a stake further out and guiding the acquisition of technology.

Call Centers and the Future of the Device

Many companies are considering ways to transform their enterprises into customer interaction centers capable of integrating all of the various emerging customer "touch points." Companies that are the most competitive are those that will take a consumer-centric approach by developing consistent points of entry for customer interaction that is consistent and personalized. For most enterprises, implementing customer caller applications where the "touch point" presents a unique opportunity to optimize the consumer request to service value and cost structure. This is the critical challenge for call center managers and executives responsible for service, sales and marketing.

The telephone is suitable for "once and done" transactions that require little or previous or subsequent transaction context. Often the goal is for each telephone call to be a single simple transaction. Complex transactions, by their nature, require the extensibility of multimedia interaction; when a multi-call transaction requires a significant amount of customer-owned information, it is emerging to be much more effective to collect this information via other media (email and fax, for example) than by telephone.

The endurance of the telephone comes from one source. One reason for the glacial pace of telephone innovation: if you own the infrastructure, you are not going to join the race to make it obsolete. The telephone number remains the most ubiquitous IP address on the planet and will be for some time. The biggest threat to the telecommunications world comes from the perception that packets are going to take over. When you consider the growth of wireless and the poor quality of telephony over the internet, this isn't likely to happen for some time.

70 percent of business today is still conducted over the telephone and this trend is expected to continue.

"The most common PC over the next decade will be the cellular phone," says George Gilder. This isn't so inconceivable when considering the continued growth of the phone as the medium for communications. Today there are roughly 54 million cell phones with 32,000 new subscribers added every day. By 2005, 47 percent of the North American population will have a cellular (PCS) phone, which is significantly greater than current PC penetration. 41 percent of North American households already use cordless phones.

All these phones mean more calls - not less - and these people will be calling you. Voice as a dominant medium will not be going away for some time, which, by the way, supports the confidence that we have at Nortel that the communication solutions you use in your enterprises will continue to come from the Nortels and Lucents of the world - not the Ciscos.

No form of communication has the telecommunications industry as concerned as that of internet-based telephony. Have no fear, IP telephony is one of those internet applications that is getting far more attention than the market is supporting. In fact, the number of internet telephony servers with PBX/ACD systems in call center installations is still small, probably less than one hundred to date.

Internet-based telephony requires the use of a desktop PC or some other undetermined device. Eventually this process will be replaced by a device as common, easy and convenient as the telephone. The current telephone offers customers and users everything that the internet-based telephone does not.

We can say with some certainty that IP-based telephony will emerge as an important transport. Keep in mind that these are still calls and there will still need to be agents and call centers to take them. For now, there are some evolutionary factors that are driving the overestimated impact of internet-based telephony: 99 percent of the population does not have a static IP address; 99 percent does have a phone number.

Internet telephony, if it is to take off, will first have to become as easy to use, convenient, accessible and reliable as the telephone. There is no killer app for IP telephony identified yet. IP telephony is still complex, of questionable quality and does not save enough money. The current state of IP telephony is where the PSTN was in the early 70's before the advent of signaling systems that led to intelligent services, such as call forwarding and number portability. Circuit switches (such as PBXs) are one-tenth the price of IP telephony switches. The transmission cost of circuit-based phone calls (i.e., long distance) has become very cheap; two cents a minute for long distance when you strip away the tariffs and taxes surrounding those long distance calls.

To make desktop call control work requires PCs that are equipped for voice transmission; in a sense they have to be designed like the telephone. Although the PC should make a good telephone interface, the fact is that there is a certain comfort level users have come to expect in their experience when using the telephone.

Eventually the desktop phone in its current 12 button configuration will go away, but at least for the time being we can't force companies to reroute their voice traffic over their networks, because many companies are not ready for that next step.

When enterprises start looking at plans to come up with a PC-based phone, they get shot down pretty quickly because the reality is that management, much less than call center agents themselves, doesn't trust their computers. They don't trust their LANs, their servers still go down quite frequently and they can't get Microsoft's Outlook working right and so on. Everything that is dealt with in a corporate environment and the prospect of having the phone tied up like computing is scary.

The telephone is the cockroach of the computing world, though it is slowly being evolved into the PC. The fact of the matter is that you couldn't kill the telephone network with an atom bomb if you wanted to. But just because you can't kill the internet with an atom bomb doesn't mean that it won't be killed by the lunacy of major corporations. The world is now made up of a fifty year network of enormous complexity that presents a simple common interface to billions of users worldwide.

The Demographics of Contact

We rarely consider the influence of demographics on call centers. The impact is more profound than given credit. When we try to understand the loyalty of customers, it is (or will become) important to understand who they are, where they live, what kinds of resources they have and how they are changing over time.

If we could shrink the world's population to precisely 100 people, with all existing human ratios remaining the same, these one hundred people would look like this: there would be 57 Asians, 21 Europeans, 14 from the Western hemisphere, including North and South America, and eight from Africa. 51 would be female and 49 would be male. 70 would be non-white and 30 would be white. 66 would be non-Christian and 33 Christian. 80 would be living in substandard housing. 70 would be unable to read; half would suffer from malnutrition; one would be near death and one would be near birth. Only one would have a college education. One-half of this village's entire wealth would be in the hands of only six people and all six of those would be citizens of the US.

One of the most significant, underestimated influences on enterprise will be the age of our customers. When we start developing customer retention strategies based on the "lifetime" value of customers, knowing how long they are going to live becomes important. In 1960 there were

5,000 Americans who were 100 years old or older. Today, there are more than a million and there will be 5 million by the year 2010.

Not only do we have to understand who they are, where they are and how old they are, but we have to understand what kinds of devices and interfaces our customers will be using. Martha Rogers, the co-author of the best selling books, The One to One Future and Enterprise One to One, defines the importance of understanding what our customers use to talk to us with by expressing that if your customers live in the ghetto – where TV penetration is higher than telephone penetration – then you need to communicate with them through their televisions. (There are more TVs in the US than there are flush toilets!)

The telephone is a significant device, but not as pervasive as we would want to think. The stark fact remains that 50 percent of the world has never made a phone call. Many are often surprised to find that 30 percent of American and 35 percent of Canadian households are not able to use DTMF - dependant call processing. These callers are restricted to dialing in pulse because one or more components that make up the phone call – CO switches, CO lines and handsets – can't handle DTMF signals. Amazingly, 12 percent of the population is still rotary.

What is evolving here is the phone, in fact, roughly one-third of all PBXs are now shipped with ACD/MIS and account for almost all of the CTI shipments.

Toll free dialing has been around since 1967 and the 1-800 toll free service as we know it has been around since 1980. Americans dialed 20.6 billion calls in 1997; that was more than 56 million calls per day. It took twenty-nine years to exhaust the 7.7 million toll free 800 numbers. The rate at which companies are reserving the toll free 888 number is 62,500 per week; at this rate it will only take 2.5 years to exhaust the 888 prefix.

While electronic commerce gets all of the press, dependency on the PSTN is growing. We are adding an average of three new area codes every six weeks.

When it comes to the call center and its economic impact alone. Call centers are even bigger than the long distance services that fuel them. In North America, we currently make an average of 60 million calls each day to 800 numbers alone. Speaking of cultural pragmatism, despite the advent of electronic banking, we still write over 60 billion paper checks per year.

Telephone penetration is rising in all parts of the world regardless of income level, as the cost of installing telecommunications infrastructure drops dramatically. Lower income countries are registering double the growth rate of high income nations in newer wireless services as they attempt to leapfrog landline technologies.

A US Commerce study on computer and telecommunications sales found that from 1990 to 1996, sales grew 57 percent to $866 billion. Significant dollars are being spent on information technology. $54 billion market in the US for carbonated soft drink market. Average number of eight ounce servings that each American consumes a year. This is versus the internet revenues of around $1 billion in 1997. The world's revenue from telecommunications services is $600 billion, that's billion. The US is responsible for $179 billion of that.

Call Centers and the Internet

The internet is an unnatural phenomenon happening at quite a natural rate. When we talk about the internet and its effect on call centers, we have to look at it from the perspective of how the internet will displace the primary interface of the call center, the telephone. The Gartner Group predicts that within several years of the turn of the century, 35 percent of all transactions will be electronic (internet). This is certainly amazing growth, but it still leaves 65 percent something else (chiefly the telephone).

The growth of the internet and the migration of services and applications to it is consistent with earlier technology curves. We should not confuse explosive growth in traffic with a quickening of the cycles of technology. Likewise, we should not confuse a quickening in the number of software releases with the quickening of long term trends. The current trend of technology enabling customer interactions has a long history; we are just at the point where it is much more practical. We have been trying to make an interactive world for 35 years or more.

Up until now, and probably for a good time into the future, the internet has been a phenomenal waste of money. And here are four oxymorons that belong together: "soft rock," "military intelligence", "taped live" and "internet advertising". Some predict, and I agree, that by the year 2000 internet banner ads will be gonc; they just don't work and they are pushed out of our periphery the same as roadside billboards are. Banner burnout. And sure, most of the growth in phone numbers is for fax and data modems but even the internet has its limits. Vinton Cerf, founder of the internet, predicts that the internet will run out of IP address by 2003.

Talk about internet usage rates. In California (a highly connected state) a typical ISP supports 15 end users per line. Generating on average a 3 percent usage basis per week, compared to 4.4 hours per day of television usage, web usage has a long way to go before people spend as much time on the net as they do in front of their televisions!

Even though there is still plenty of room for growth in telephony globally, the 80 million global web users is a small number compared to the world's 800 million phones and 6 billion people. Of those 80 million, 56.4 million of them are in the US, next is Japan with 8 million, the UK with 5.8 million, Canada with 4.3 million and Germany with 4.0 million. If these are not your customers, then the phone is still likely what they are using.

Even when we talk about a totally digitally wired world, The Gartner Group estimates that by the year 2002, 65 percent of "connected" users will still be using analog modems for their main access method. Most customers (as much as 70 percent) only have a single phone line in their homes. About a quarter of American households (which is less than 2 percent of the world's population) have computers with modems. 15 percent are expected to be ISDN based, 10 percent cable modem, 5 percent wireless satellite and 5 percent using Digital Subscriber Lines.

When we look at the end points of consumer technology, this is where we can find the pragmatic grip of the pace of technology. More than one-third of North American households do not own a PC and have no intention of buying one. Forrester research analyst Ken Clemmer is quoted as saying that "Generally these people are older; they are technology pessimists. Most just don't understand how a computer could make their lives any better. There is no killer application for their group." More than half of non-PC owners are over 65 and nearly 80 percent say they earn less than $35,000 a year. 84 percent had no need for PCs, while only 6 percent cited cost as the main reason for not wanting one.

Referring earlier to the influence of demographics on the acceptance and use of technology, there is a tremendous inequity in the "balance of access." Knowing exactly who your customers are influences an understanding of how they will communicate with you.

The number of non-white households in the US without a computer and internet access speaks for itself. In fact, while the media glamorizes a fully wired world, looking at the dramatic and deplorable statistics of school connectivity in our ghettos and inner cities casts a pall on the digital future. Unlike the telephone, those that have access to the internet compared to those that do not demonstrates the existence of an increasing gulf between the information haves and the digital have-nots.

But, where there is money to be made, investment will lead. The amount that Americans spent on retail goods in 1997 was $2.5 trillion. Even if web-based retailing captures just 5 percent of all retailing revenue over the next ten years, this would be a steady growth of $125 billion a year. With over 20 percent of US households online, revenue from shopping on the internet exceeds a billion dollars a year and is growing rapidly.

The Future of the Internet

The Internet is important - after all it will be supporting one-third of all electronic commerce before long, according to The Gartner Group.

This new generation of contact centers can field customer inquiries via any communications medium and respond via the same or any other communications medium.

The economic benefit of using the internet for certain types of transactions is very compelling. In fact, in some applications internet transactions can cost approximately one-tenth the cost of live telephone transactions. Web support not only reduces the number and length of calls, but advanced internet integrated computer telephony applications provide sharper skills matching and call routing between callers and support personnel. Electronic support is all about empowering end users and customers to solve their own problems.

What we are starting to see, though, is a parallel between the evolution of functional web sites and the introduction of IVR (VRU) several years ago. As call centers began to realize the benefits of IVR, many call centers and enterprises began to deploy numerous self service applications in a stand alone role, but had to learn quickly to design applications that provided callers the option of "bailing out" and talking to a live agent or risk the accusation of being trapped in "voice mail jail". In the short history of the self service web application, this bailout is once again beginning to surface, becoming critical and creating the need to fully integrate the web with call centers via CTI technology to support this integration. Web-based self service will not address the need for live service, there will always be web users who will prefer to speak to live agents, just as with IVR.

It is more compelling than IVR was, but has the same kind of "burn-in."

The most important influence of the internet may not come from its ability to serve as a mixed media catalog (brochure ware), but in the influence of propagating email. One of the "killer applications" of the net is email and its importance is growing.

With all of the microchip muscle backing it up, email may actually be a cultural phenomenon of its own in contributing to the revival (survival) of the oldest of the office tools (writing). 50 years ago, people vociferously complained that because of telephones we were going to lose the art of writing. Email may actually be teaching us and preserving the art of expressing ideas using skills that we may not have used five years ago – although the annual shipments of stationery are running 15 percent less than a decade ago.

15 percent of the US population now has access to email and this will grow to 50 percent by 2001, with North Americans sending some 500 million messages annually - it's that blistering pace again. What's important is that the use of email for customer contact will grow from minimal today to 10 percent of all customer contacts by 2001. These email messages, because of the increase in delivery reliability (remember the AOL outages), are coming with a response expectation that is no different than customers' speed to answer expectation. Today, 95 percent of all messages delivered through the internet are received within five minutes, up from 81 percent in 1997.

In an age where employees, agents and customer service chain executives are squeezed for time and bombarded by information, email has its own advantages, disadvantages and pitfalls as well as its own unique idiosyncrasies. The subtleties of email are different than, and make it difficult to pick up on the subtleties of conversation, compared to the telephone, where an agent can hear the reactions of the other person and adapt the nuances of the conversation. With email, the sender's intentions are abstracted another step further.

Email adds a new media and twist to customer service politics. We have heard the complaint about voice mail jail and not being able to reach a live person so often now that we ignore it. Email can be worse. If the etiquette of email is emerging as being fine for quick messages, but not for condolences or most thank yous, then we should balance this against the use of email for sensitive customer issues.

It is still impressive to receive a thank you note in the mail from the companies we do business with. Because it is becoming more rare, from a customer enterprise relationship perspective, it is more unexpected and hence, more personal. Sending a written note should be part of the savvy enterprise's workflow process. The very nature of the ability for us to distinguish the quality between an email contact and a written note, should serve as notice that email, for the time being, cannot replace (type)written letters and communications.

With the exception of those transactions where speed is the quality criteria, not every customer who uses email wants better or instantaneous collaboration. Avoiding people is an excellent use of email. The customer has the power to make thoughtful, non-verbal choices about the communication, which more than anything, leads people to use email.

The Future of Everything Else Customer

What credible researchers are forecasting is that there will be a progression or transition of acceptance and usage of the various interfaces, mediums and devices influencing the call center. This progression started some time ago with mail. Today, for the most part, as a customer service tool, the mail is definitely gone – nobody writes the customer service center anymore. In the early 70's and 80's, you would just write a letter saying that you needed help and wait for their help. Next came the fax, a ubiquitous gift from the Japanese who have too many letters in their alphabet – kanji – to put all on a keyboard, so they designed a technology to take pictures.

Next came the 800 number, which we will see taper off by the year 2010, according to research by the University of Indiana.

What will be pervasive is internet-based fax and internet telephony. It is starting to take off now and will outgrow any other channel that we will see in the next ten to twenty years – this of course is the Web. IP telephony will emerge as the dominant voice carrier. The growth of reliability and quality will inevitably surpass these legitimate issues we have today. What this research says is that IP will have its greatest influence over the customer-to- business contact

All of the new emerging networks will be built on two basic logical building blocks, a core transport/switching layer and an access edge. At first, the core transport/switching infrastructure will be based on either IP or ATM core switches.

The future in consumer products is in the products that blend technologies (converged technologies) TVs and PCs, fax, scanners and copiers in one. It is estimated that by the year 2002, a quarter of US companies will adopt multimedia technology into their call centers, according to Datamonitor. Datamonitor forecasts that in 1998, web sites geared towards e-commerce will receive inquiries from 11 percent of their non-casual visitors. Corporate sites will get about a two percent email rate. Datamonitor, US multimedia Call Centers.

The popular media even influences call centers. One hundred percent buzz-word compliant.

We don't change our interfaces as quickly as people and the popular media would have us believe. If you count the growth of wireless phones (and why wouldn't you?), it is clear that the telephone, specifically audio as a form of communication, is not going to go away for some time (perhaps as many as twenty years or more).

While that 35 percent of the pie gets 98 percent of the media attention (how many telephony magazines can you find in the Borders or Barnes & Noble magazine rack?), it will remain the telephone (and your job) that will handle the bulk of customer interactions for some time to come.

When you stand at the Barnes and Noble magazine rack, you see that 95 percent of the magazines are about the internet and PCs; nothing is about telephony except for the stray Teleconnect or Call Center Magazine. Or a computer telephony. There is a disproportionate amount of attention spent on "other" media, not that this is not important. If one percent of your customers use these media, you have to provide it to them, in the customer loyalty model and the downstream effect.

Many believe that some sort of revolution will happen to get rid of all of the technology found in call centers and drive the enterprise back to just answering calls (or in the case of the information age, whatever media a "call" is). Back to a point when the call center agent was sitting at a desk and knew exactly what the problem was and could answer every question, in every satisfying way.

Conclusion

Call centers are, without question, not at the heart, but are the very heart-beat of today's customer focused enterprises. With the exception of a store-front counter, the enterprise's call center is about as close as the enterprise can get to its customers.

If Peter Drucker is right and most of the information (re)processed by the enterprise is internal, then call centers serve as one of the few sources of true and real time information that an enterprise has. This makes call centers the battery farms of the information age. Nobody wants to talk about batteries, they are just not as sexy as the internet, multi media and virtual reality. Nevertheless, thanks to the increasing pervasiveness of technology, customers want to get closer in many more ways than ever, meaning that there are many more "battery types" than ever, each with its own set of expectations and protocols.

Customer interactions with the call center and all of the other types of self help applications and mediums such as the internet, email and IVR now make a customer's entire impression of the enterprise. These impressions are less and less influenced by mass media marketing and more and more by the ability to recognize customers, their needs and preferences, and deliver real time solutions.

Call centers are where a customer's "moment of truth" happens with more clarity than with few, if any, other engagements with the enterprise. In that "moment", if your customer's needs are not met expeditiously and professionally, you stand to sacrifice quality in your relationship. One of the mistakes enterprises make today in the employment of technology to cultivate relationships is reducing points of human contact; unfortunately the enterprise sacrifices valuable opportunities to cultivate the very relationships they profess to value so much.

Those enterprises that have grasped the opportunities to exploit the cultivation of life long relationships with customers have taken advantage of the advances in the convergence between computers and telephony to move their call centers from the corporate backwater to the corporate backbone.

Wal-Mart, USAA, Nucor, Southwest Airlines and Home Depot are examples of companies that sell the same things as their competitors: supplies, tools and insurance. What makes them different is that Wal-Mart, USAA, Nucor, Southwest Airlines and Home Depot are also examples of companies that are able to prosper by satisfying customer needs that are being ignored by existing competitors who are mostly focused on beating each other up. It is the competitor that you can not see or hear, in the case of call centers, that will kill you.

We are also at a point in time where we have to give some pragmatic consideration and look at the true effect the application of technology has on customer service. Call centers are one of the few points where an enterprise has a human face, or at least a voice, anymore. Now at the confluence of critical commercial interactions between business, prospects and existing customers, call centers play three key roles: enhancing customer sales and retention, serving as a front line of support for internal and external customers and as the focal point for service operations.

Technology rarely is a good substitute for good business practices and never a very good substitute for superior customer service. Given the choice, most customers will choose superior customer service over things such as quality products.

What most customer service chain executives have been influenced by is the belief of expecting the call centers to be the magic bullet or total solution to a company's communication problems with its customers. In the relatively brief history of customer service technology, we know that an enterprise cannot just install an ACD, IVR or CTI and leave it at that. There are many large companies today that don't understand that the technical interface between the enterprise and the customer often simply replicates the poor ways of already doing things in the first place. "If we keep on doin' things we always done, we'll keep on gettin' what we always got" (Barbara Lyons).

Today we are so often focused on the technology that we often forget to ask if we are really serving our customers. There is still a sizable population of customer service executives enjoying satisfaction at the miserably high failure rate of technology-based applications in call centers (mostly because technology is competing for their jobs). You would never get any of these executives to admit it, but it should be of no surprise.

When looking at the speed of technology's change over the past decade – the demographics, interfaces and devices customers used – it becomes evident that the movement from a phone-centric model of support to an electronic one will not happen at the speed we want it to, nor will it go in the direction we expect. In fact, the fallacy of picking one technology over another is just that what is actually happening will not be the emerging dominance or distinction of any particular technology, such as the internet, but less and less a distinction between emerging technologies.

The enterprises that are successful will create and market services with the same care and strategic analysis that they devote to their product offerings. When talking about the next generation call center, it is virtual, it's technology enabled, it's out-sourced integration, it's standards based and customer centric.

Leading edge call centers will have most, if not all, of these characteristics. Using technology basically as a competitive weapon against their competitors, call centers are taking advantage of the benefits of:

The ability to lower cost without the lower quality or access
Total customer satisfaction by any metric
Member attraction and retention
Revenue growth
Market differentiation
Higher fees for superior service
Decreased variation across the system
Increased customer expectations

The successful innovators will be those that develop the best understanding of the underlying change and act upon it. Success in the next ten years will depend on and require a deep understanding of the change in the power of interactive and communications capabilities in not just call centers and the enterprise, but in the global economy in general.

Service in the mixed media environment is not the same as in the phone-centric environment and to survive and remain competitive there are a lot of questions to ask. It's still early in the game; while promising products are coming into the market that offer new ways to communicate with voice, you are being cautious. Most agree that the tidal surge is still three years away, which isn't so far after all.

Establishing an interactive experience, regardless of media (including live agents), that is characterized by an obsession for reliability, speed and intimacy is the most significant way to effect and forge the relationships that will characterize the successful enterprise of the next century. The expectations of speed, reliability and intimacy are increasingly becoming the "expectation benchmarks" that your customers are using to measure the quality of the interaction with your enterprise. When we go look at the call center in the year 2000, the implementation of technology is used specifically to say that the enterprise can get above the bar much higher by using the web, email, fax or some other medium. These enabling technologies are the things, tools, processes and applications that allow us, in a very cost efficient way to exceed customer expectations, while at the same time internally maintaining a profit center that is meeting its goals as well.

Your enterprise must not only be experts, but futurists also. You are either right on, too early or too late. The transition of the call center from just a dumb interface to an point of intelligent collaboration. Call centers are the most visible opportunity for enterprises to meet competitive demands from customers for attention and service.

As the speed of the information age accelerates, the difference between finding a customer, growing a customer and keeping a customer is blurring. The call center of the immediate future will be charged with not only improving customer service value and cost of providing it, but also reducing the cost of attracting, retaining and securing customers – for life.

Chapter Three

A Paradigm Shift: New Call Center Strategies for Web Callers

Introduction

By the time you read this, you may already be familiar with the following kinds of statistics about the rapidly growing use of personal computers, the Internet, and the World Wide Web. What you won't necessarily know yet is how such activities should fit into your call center operations.

• According to the Spring 1998 Cyber Stats report from Mediamark Research Inc.:

- 35% more U.S. households own PCs than three years ago.

- Approximately 44 million U.S. adults - 23% of the adult population- currently use the Internet, a 260% increase in six years.

- The number of adults with access to the Internet represents 32% of the U.S. adult (age 18+) population, or some 62.3 million people.

- At-home usage now exceeds the workplace by 26%. Some 27.6 million people are estimated to have accessed the Internet from home versus 20.4 million from work.

• Based on a new FamilyPC reader survey, 83% of families polled were online in 1997 - up from 50% the previous year. The majority (59%) connects to the Web roughly 35 times a month and spends an average of 16 hours online per week.

• "A poll commissioned by the Information Technology Association of America, a technology trade group, showed that more than one in ten Americans is purchasing goods and services online. The ITAA telephone poll of 1,000 U.S. adults, conducted by polling firm Wirthlin Worldwide for the trade group, found that 15 percent of those polled have made online purchases. And buyers tend to have comfortable incomes, with those earning $60,000 a year or more twice as likely to buy online, the survey found." - ZDNet News

• "Software is the No. 1 online seller, representing 38 percent of all E-commerce, with books following at 28 percent, and PC hardware at 20 percent, according to the results of a randomly conducted telephone survey of 500 U.S. internet users and 500 nonusers published in The Consumer Online Commerce Report by Organic ... Industries that saw a rise in their online sales during 1997 were general gifts at 18 percent, automotive at 15 percent, apparel at 14 percent, and food purchases at 4 percent...The number of online shoppers rose to 11.2 million in 1997, up from 6.6 million the previous year...Seventy-six percent of online shoppers made multiple purchases in 1997, totaling $3.3 billion in consumer spending...In 1998, this figure will grow to an expected 15.5 million purchasers spending a total of $6.2 billion... The diminishing fear of using credit cards online may be a contributing factor to the rising number of online sales...Two out of three online shoppers used their credit cards online to complete transactions and more than 7.3 million users made credit card purchases online in 1997." - News report from InfoWorld Electric

What are all these statistics supposed to prove? That the Internet and the World Wide Web are slowly but surely taking over many of the communication activities that the traditional telephone network serviced. Although we are not getting rid of the telephone and voice communications, the call center's role can no longer be confined to assisting traditional telephone callers. The consequences of that statement are not trivial!

Web Self-service: Greater Convenience, Lower Costs vs. Revenue Generation

The World Wide Web is exploding self-service, multimedia information dissemination, and electronic business (e-business), while the Internet's email facilities are changing public access, two-way, person-to-person messaging. The public data network (internet) is even usurping transport and switching of voice communications, with Voice over internet Protocol (VoIP) gateways that are becoming incorporated into traditional PSTN network switching platforms. This has already begun to impact call center technologies, as Web sites provide information and invite convenient direct communications contact from both new sales prospects and existing customers via this new method of access. In this chapter, we examine the implications of Web commerce and the new contact and response alternatives for the technology, staffing, and management of Web-enabled call centers.

Personal computers, "internet appliances", and Web access will provide a new form of customer interaction for many of the things the telephone network was traditionally exploited for, including both self-service e-business and customer services. According to a recent report published by the U.S. Department of Commerce, The Emerging Digital Economy, consumers find the convenience of self-service "as the number one reason for making a purchase online."

From the enterprise perspective, there is a huge amount of cost savings to be realized from any form of self-service. The Department of Commerce report cites a Booz-Allen & Hamilton study showing telephone banking transaction costs potentially reduced as much as 98% with online (Web-based) self-services and insurance customer self-service over the Web having a cost advantage of 58-71% over telephone call centers. The report includes other case studies of e-business, showing significant enterprise cost savings from the deployment of Web-based self-service applications that would have required live assistance in a telephone call center.

Web-based self-service is also significantly changing the middle-person role in many kinds of traditionally local service businesses, including travel, real estate, insurance, and car buying. It is most certainly usurping the role of mail order businesses such as books, computer hardware and software, pharmaceuticals, and other commodity catalog sales.

However, completely automated self-service transactions and information access are not always adequate for every customer need, and human assistance will always be required for complex situations, particularly where problem resolution and decision-making are involved. Probably a more strategic consideration exists where the opportunity to have a realtime conversation with a customer is a valuable marketing opportunity for revenue generation that will often be more important than any cost reduction. In either case, that's where the next generation of Web-enabled call centers will fit into the future of e-business on the World Wide Web. Though traditional telephone and voice communications will remain essential elements of call center operations, there is already a slew of new operational requirements that will arise from Web-based access and internet-based person-to-person communications. It won't be a matter of choice, but competitive necessity that will drive businesses to bite the bullet of call center Web support.

Blending World Wide Web Self-service and Call Center Assistance

The ubiquitous telephone set made things simple for the call center, since all we needed to do was establish a voice connection between the caller and the call center staff. With IVR self-service applications, we needed to have the caller equipped with touch-tone telephones. Things become more complicated for Web-based callers, because of the variability in PC configurations that the caller may be using. Depending upon the availability of modem speeds, sound/video cards, adequate memory for multimedia client software, etc., there will be limitations on Web-based functions that the user can exploit, and, consequently, the interfaces with the call center. But even there, things are changing!

The growth of the Internet and the World Wide Web can be ascribed to the proliferation of PCs. However, the paradigm has shifted. Now that the power of the Web and the internet for information access and personal communications has been proven and accepted, "internet appliances" are being developed to exploit network usage with more simplified devices for the consumer public, rather than general-purpose and complicated "computers".

In a recent report, IDC projected the end of the PC's domination of end-user access as a terminal device. "The era of PC dominance may be nearing an end, as the explosive growth in users - and uses - of the Internet expands device requirements well beyond the design point of the general-purpose PC .... The growth of the internet is driving the development of a dizzying array of new user access devices, including TV set-top boxes, Web-enabled telephones, Web-enabled personal digital assistants (PDAs), and Web-enabled video game consoles. IDC forecasts that these non-PC devices will account for almost 50% of unit shipments by 2002, dramatically driving down the PC's share of the market."

The type of access devices employed for Web access will have an obvious impact on how users will be able to exploit self-service information retrieval, transactions, and communications. We have learned from our experience with IVR applications that live assistance is always required to handle telephone caller needs that can't be resolved efficiently or easily with "programmed" interactive applications. This means, then, that multi-access "Web callers" will place a new set of demands for live assistance on call center activities. Since we have only had experience with PC-based Web users, there is more learning to come, as these other, perhaps less flexible, Web-access devices enter the consumer market.

The Web site must be viewed as a principal means of public access to more efficient and flexible multimedia, self-service applications, and, like IVR telephone applications, to live assistance options as well. Call centers have evolved from the tradition of basic telephone answering and live voice assistance; these facilities will continue to be important even in the new era of the Web and the Internet. However, it is time to rethink the definition of multimedia live assistance and communication responsiveness that will evolve from both World Wide Web access and the rapidly growing availability of universal two-way email messaging. The blending of realtime Internet voice connections with messaging options is both a challenge and an opportunity for call center management to improve effectiveness of service performance and reduce costs. Most importantly, it will be a necessary, rather than optional step towards meeting the new needs and demands of the caller/customer in the next millenium.

Insight: When Charles Schwab reports that 50% of its trading activity is now done by online self-service from its Web site, you had better believe that it is the convenience and speed of such services that primarily attract their customers, not the need for call center assistance. Only when the Web caller runs into a problem, does call center assistance become necessary. The same incentive can apply to any sales and customer service support, when it becomes convenient and easier to "do it yourself."

What's A "Web Caller"?

For purposes of discussing the impact of the Internet and World Wide Web, we need to differentiate the caller/customer who accesses the enterprise through its public Web pages, instead of the telephone. Because such access will inherently enable new self-service capabilities that will change their support needs, Web callers cannot be treated simply like telephone callers; they will be also be more amenable to new response alternatives, such as two-way email messaging. It is therefore up to the call center to properly structure its response capabilities to conform to the various choices that the callers can exercise, within the context of available call center resources (staff).

Web callers are defined as people who contact and interact with enterprise call center staff through the Web page application procedures and facilities made available to them.

Insight: Although it's OK to handle a traditional telephone caller with a Web-enabled call center, it's not so OK to handle a Web caller with a support representative who doesn't have access to the same information the Web caller does!

The critical factor for Web callers is call center access via applications-controlled Web pages and structured input forms. Initial contacts with an enterprise workgroup will rely on a Web page as a starting point, even though subsequent communication interchanges may be freeform email messages or realtime connections. To highlight this emphasis on Web-based access, the term "Web caller" is more descriptive and it includes two-way messaging options to telephone calls as a form of call center communication activity. Needless to say, the same Web callers may also use traditional telephone access to the enterprise call center, so both forms of support must be available to an enterprise customer.

"Come Into My Web Page," Said the Spider to the Fly

Another new, but strategic, consideration about Web callers is that once they hit a Web page, that fact can be detected in real time, and can enable the Web server application software to dynamically react to the connection of the Web caller, even without knowing the purpose of the current access. This opens the door of opportunity for various proactive applications to be triggered, including the involvement of realtime interaction with call center staff.

Insight: The intelligence embedded in the Internet allows two-way accessibility whenever a Web caller logs on to the Internet. This is unlike the PSTN, which only lets the caller initiate a specific call destination. This new power can be used for overcoming the traditional outbound telephone problem of making contact with someone who is not accessible at a particular phone number, i.e., busy or no answer. It can also be abused by aggressive telemarketers and there will have to be safeguards for this new invasion of privacy.

Web-based interactions may include various forms of connectivity, ranging from realtime voice/chat multimedia connections and telephone callbacks, to message exchange via email and fax. Furthermore, since email offers multimedia options in terms of data attachments and Web links, email responses can also consist of text, voice, and even video information.

Finally, Web callers are not necessarily restricted to Web access, but can always revert to being traditional telephone callers, depending upon their particular needs and circumstances. However, the continuity and consistency of all forms of contact between the enterprise and the Web caller must be consolidated and preserved, regardless of the particular method used for a current contact. So, no matter what communications have taken place before, the results must all be taken into consideration for handling the present one. It's going to be a very dynamic "mix and match" world of communications media access!

Web Applications and Call Center Assistance

Call center assistance skills have always been based on upon the call-handling applications involved. Thus, the personality, verbal language skills, and subject expertise for customer service, sales, or collection applications, are the important qualifications for traditional inbound/outbound call assignment to call center staff. The differentiation of such call center applications becomes even more important for the strategic use of call center involvement with enterprise Web sites, since it not only affects skills-base routing (SBR) strategies, but also whether the Web application should initiate a realtime live voice connection (via internet telephony), a text chat connection, or callback/messaging communications.

It will make a big difference to the call center access strategy if the Web caller is a "surfer" (a first-time visitor), or a frequent visitor/customer at a sales Web page. It also will be a different story if the Web page is devoted to customer service support needs. Like using Dialed Number Information Service (DNIS) and deploying different 800 numbers for sales or service telephone access, the various application-oriented Web pages will trigger different user interfaces and different response options. Furthermore, like Automatic Number Identification (ANI) for inbound telephone call handling, Web "cookie" information stored on the Web caller's PC can also invoke personalized information displays (a caller's "screen pop"), and personalized response options.

How The Web Caller Will Impact the Call Center - The Top-Down View

The Web caller has to be viewed as different from the traditional telephone caller in some ways, while similar in other ways, but with new options for satisfying their needs. Such differences, however, mean that all the rules of the call center game have to be adjusted, from performance metrics to resource management tools and strategies, in order to accommodate Web callers effectively.

The best way to assess the impact of Web caller access to the call center is to look at the "users" and the human interfaces in the call center operation. I break down such "users" into three fundamental categories:

1. Callers /Web Callers - These will be customers or prospects who initiate contact to the call center from a traditional telephone, a Web page, or who may be contacted by call center staff on an outbound, callback basis.

2. Web Support Representatives (WSRs) - Often referred to as "agents" from the old days of reservation systems, CSRs (Customer Service Representatives), or Telephone Service Representatives (TSRs), the call center representative is anyone whose primary responsibility is to communicate directly with callers/customers by realtime voice connections (inbound/outbound) or by messaging. To highlight the new responsibilities, skills, and interfaces that will be required to handle the Web

caller, it is more than fitting that we give call center staff that will support the Web caller a more descriptive label, such as "Web Service Representatives" (WSRs).

We may have to break down this classification even further, to accommodate the significant differences in message response support from realtime voice (telephone) interaction. This differentiation will help in the planning and management of staffing resources as Web caller activities become well-defined metrics.

3. Management - All areas of operational management will be affected in the ripple effect of handling the Web caller. This will mean keeping the old responsibilities for telephone callers, while adding new Web caller responsibilities as well. Such management responsibilities are primarily focused on the planning and control of both staffing resources and call center application technologies, processes, and procedures to cost effectively satisfy caller needs.

New call center technologies and the Internet are making call center capabilities much more flexible for both traditional callers and Web callers. It is important to understand and address what the callers will really need or want from call center support, both technology-wise and skill-wise from call-handling staffs.

The Web Caller's Perspective - Different From the Telephone Caller?

It is critical to start from the top with the Web caller, because it is this new breed of call center contact that must be fully understood in order to identify what must change in traditional call center operational functions. Even though experience with Web callers is still evolving in the footsteps of new technology deployment by leading-edge call centers, there is enough known about traditional telephone activities that can be logically extrapolated to the new Web caller and internet messaging environments.

How will the Web caller be different from the telephone caller? Here we must look at the subtle situational and psychological differences that can affect the caller's support needs. The expectations of the Web caller in terms of speed of response, information access, and technological preparedness, are all dimensions of interaction with the enterprise call center that must be evaluated carefully.

With the traditional telephone call center model, callers have instant dial tone before placing a phone call, expect no busy signals and a short number of rings before being connected, and then experience either an automated self-service application or a minimal wait for live assistance. The pain of waiting in a call center queue for assistance is exacerbated by the fact that telephone callers are truly being forced to waste their time waiting on hold. That's why speed of (live) answer has been a traditional metric for quality of service. The caller's alternatives to waiting in queue for live assistance include: abandoning the call (hang up), using a self-service option (that is often not too useful for the caller's real needs), or leaving a voice message for a callback (scheduled or ASAP).

Self-service for the telephone caller, using IVR applications, has always proven very effective for certain kinds of applications, such as simple (and short) information retrieval and limited data entry transactions. In fact, the Gartner Group estimates that self-service transaction costs are between 15-20% of the costs using a live agent. When 40-80% of the telephone callers can be completely serviced by the IVR application, the payoff is truly significant. However, for those applications that are too complex and cumbersome for sequential interactive menus or require textual data input, live assistance from call center staff is always necessary. Now the limitations of the telephone and voice can be bypassed and the potential of self-service expanded dramatically for Web callers and their screen-based PCs.

Self-service Applications, Not Dial Tone, Are the Starting Point for the Web Call

For the Web caller, the circumstances are somewhat different. The fact that the Web call originates from a screen-based Web page means that the Web caller is not navigating time-consuming (and often frustrating) audio application menus. Unlike complex interactive voice response interfaces, screen-based applications are inherently more efficient and manageable because of random and concurrent access to menus, help information, and other kinds of information that may be useful to the Web caller. It is not only faster, but facilitates efficient caller "multitasking" (doing several things at the same time). For this reason, call center self-service applications that used to rely on voice-only IVR technology, will find greater flexibility and effectiveness in the screen-based Web server environment, including access to live assistance.

Although today's Web caller uses a PC, the expansion of Web access through a variety of visually-oriented "internet appliances", including TV set-top devices and new, "smart" screenphones, will also bring practical advantages over the traditional audio-only telephone access to the enterprise call center.

Self-service applications associated with the call center must still be consistent for both voice or screen interaction by callers, whether accessed by telephone or Web. This is important for application design, implementation, maintenance, subsequent interaction with call center staff, and to minimize any confusion when callers switch from one mode of access to the other. This requirement has been recognized by the leading IVR software vendors, who have expanded their application development software tools ("application generators") to include both voice and screen interactive user interfaces for the support of common application call flow logic and infrastructure integrations (application database servers, operating systems). This means, that self-service applications can and should be designed for both telephone callers and Web callers, including access to live assistance and messaging options for both. However, there will always be differences between the two, in terms of caller options and interaction logic, because of the inherent nature of the voice and screen interfaces.

The Personalized Screen Interface for Web Callers

There's always been a lot of talk but little action for truly personalizing the caller's interface with a call center application. About the best that has been done within the limitations of ANI or an IVR caller identification input has been to activate the CSR's terminal "screen pop" for a particular application, along with database information about the caller. With self-service IVR applications, it is quite tedious to design and implement an interactive voice script that can be personalized, by using text-to-speech technologies for voicing the caller's name. With screen output, however, the personalized text is easily incorporated into the Web caller's interface, making Web-based applications more customizable, i.e., a "screen pop" for the Web caller's desktop. Screen output permits more information to be given to the Web caller than is practical for audio output to the telephone caller.

The more successful e-business companies have exploited such personalization quite effectively. For example, Amazon.com greets visiting customers by name, gives them access to their account information, and proactively notifies them by email of book availability (based on purchase history or specific requests). Given the past history of customer preferences, e-commerce Web site applications can automatically tailor the Web site information presentation and options to match customer preferences, or base them on currently outstanding transactions such as order status or billing problems.

Another aspect of personalization can involve enterprise support personnel that have specific responsibility for a customer's particular needs, like an account sales rep. Since Web caller intercommunication does not always have to be a realtime conversation, two-way email exchange with a call center workgroup (or an individual representative) can be a practical alternative to waiting in queue for a realtime connection. Customized person-to-person messaging, as discussed later in this chapter, will usually be more expensive to support than voice conversations, and should be reserved for high-priority situations in a call center environment.

The Web Caller and Online "Flow"

One of the phenomena experienced by people who access Web sites is a time warp called "flow," defined as a psychological state of high involvement. Donna Hoffman and Thomas Novak, marketing professors at Vanderbilt University, have been researching the behavior of Web surfers and its implications for e-business.

In an article in the Los Angeles Times, Hoffman and Novak describe flow as "what happens to you during network navigation. When you are completely focused on the activity...there is a loss of consciousness about what's happening in the external world. ...At the end of the experience, you have this sense of satisfaction..." According to Novak, "One of the most important components in our flow model is perceived control. The navigation has to be under the user's control and choice, has to be nonlinear and non-hierarchical."

The implications about flow may be subtle but important for e-business and its interaction with call center functions. On the one hand, according to Hoffman, "Flow may be important for encouraging repeat visits or repeat purchase behavior." On the other hand, what happens when the Web caller needs assistance and stops the flow by abdicating complete session control to a Web Support Representative in the call center? The transition can be disconcerting when the Web caller clicks on the "Call me" button for call center assistance.

The Web Caller's Options for Call Center Assistance

When the Web caller needs assistance beyond that provided by self-service informational or transactional Web applications, more selective options are available based upon the identity of the caller, the context of the call, and, most importantly, the caller's choice for accessing live assistance. Information about expected waiting time would be helpful to the caller in deciding whether to wait, leave a callback message, or indicate urgency. What is most important, however, is that intelligent decisions can now be made by both the caller and the call center routing software, rather than simply second-guessing what the Web caller will accept as "second prize" to an immediate live connection. The reason for this increased intelligence is very simple; the flexibility of the screen-based interface enables greater and more efficient information exchange and interactive negotiations for options between the call center system and the Web caller.

Of course, there are many situations where the option for live contact will be decided for the Web caller a priori. Telesales operations, for example, would want to have an aggressive salesperson take control of the call and engage a prospect immediately in a live conversation (voice or text chat). In such a situation, a browsing Web site visitor will be targeted for a Voice over Internet Telephony (VoIP) connection, which would enable the conversation to utilize the current, single Internet access connection. One can make the analogy here with outbound telemarketing activities, in that the Web surfer is relatively passive and the call center representative is initiating the conversation. The big difference is that the Web surfer already has a connection to the Web site and may have taken some explicit action to enable the call center connection.

Working While Waiting for Live Assistance

Waiting for assistance does not preclude productive activity on the part of the Web caller because of the multitasking capabilities of today's PC operating systems. There doesn't have to be any disconnection from the call center's Web page while the Web caller does other useful work, whether it be browsing other sites or performing some other PC application task. Then, too, Web callers are conditioned by slower Web response times, rather than immediate dial tone connections, so they will be more inclined to wait for a response than traditional phone callers. The bottom line here is that Web callers are already in an "online" information access mode that they control, and are not being forced to waste any time "holding" for assistance. So, even if they aren't immediately connected to a live person, they can essentially look at information or do email, while still maintaining their call center access connection.

Letting the Web Caller Decide!

With all the sophisticated weighting algorithms that may be employed in determining what kind of call center staffing skills will best satisfy the caller, "skills-based routing" (SBR) doesn't solve the problem of understaffing and tactical call assignments. In the simplest case, calls may be routed to anyone that is available within a certain skill group and it won't have to be a match made in heaven. However, once there is too much traffic for available resources to handle, it's going to be very difficult to make any accurate decision about assigning available resource skills. One of the benefits of having the screen interface is that it provides more options to the Web callers for assistance and support that will be acceptable for them.

Here's the way this can work: if the waiting time for the skill level that would normally be available becomes too high, rather than forcing such a wait, the Web caller can be queried for more information about the importance of their problem or about using other alternatives, including leaving a message or being directed to a FAQ knowledge database. If the Web caller just wants to discuss something with a warm body ("warmware"), or has a simple question that anyone can answer, that fact will make the call routing process a little more "intelligent." If the Web caller needs to talk to someone in charge, that is obviously a clue to upgrade the priority of the call, leave a message, or to warn of a longer wait for connection to assistance.

Insight: Just remember that waiting on the Internet doesn't bring with it the added penalty of paying for telephone charges that is becoming more commonplace for technical assistance.

Immediate Call Center Connections - "Callthrough" vs. "Callback"

It should be obvious from previous discussions that the power of call center assistance to a Web caller stems from the ability to share the same Web page informational content and for the Web Support Representative (WSR) to consultatively guide or control the delivery of such information to the Web caller's PC screen. It is also important for the Web caller to have such dual connectivity in order to enable continued utilization of Internet facilities while waiting for assistance. This includes being accessible for realtime telephone calls, even while busy on the Web; Central Office telephone services will be offering "Internet call waiting" service options for this purpose.

The first attempts at providing call center access to Web callers required Web callers to have a second telephone line for voice callback, while retaining the first Internet connection for the Web page. Instead of a separate second line, a higher bandwidth ISDN connection would also work. However, internet telephony "gateway" technologies now allow a single Web caller phone line to be used concurrently for both the Web page and the internet telephony voice connections. This technology can be used for both an immediate connection with traditional call center system, or, with "Internet call waiting" service now available from telco

Central Offices, for callbacks to Web callers with only a single telephone line. The call center's Web application server can detect the Web caller's PC hardware/software configuration and download appropriate client software or Java applets for establishing the Internet telephony voice connection. (Web callers will still need a fast modem, sound card, and speakers for using the PC as telephone!)

Insight: Even though computer telephony will move call control features to the desktop screen, the traditional telephone handset (or new screenphones) will probably remain the device of choice for Web callers, rather than headsets and microphones.

Web Caller Messaging - "Web Mail"

If Web callers decide not to wait for immediate live assistance, organized messaging is an option they can exploit for a scheduled return telephone call or a message response. The kind of Web caller messaging we are talking about will be structured, i.e., Web forms, and contain the necessary context information that can facilitate automated message distribution and response processing. Because Web callers will implicitly have personal email mailboxes, two-way messaging is now a more viable alternative to live telephone connections. Not only is there public access to the Web caller's personal mailbox, but multimedia informational attachments that might normally be mailed or faxed can also be immediately emailed digitally in response.

Once a messaging transaction has been initiated between a Web caller and the call center, when a request has been processed and a messaging response delivered to the Web caller's mailbox, there is no guarantee that further communication with the Web caller will not be required. This presents a complication to Web mail processing that is analogous to a caller having to make more than one call to resolve an issue, and there is now an existing context that must be included for efficient follow-up contacts. In such cases, the Web caller may want to talk to someone immediately, or reply to the call (Web) center's response message by normal (freeform) email. A message transaction tracking number ("trouble ticket") will be important for all such follow-on correspondence from the Web caller. However, because such messages will involve freeform email, Webmail processing technology products must include scanning through such messages as well, to insure proper content identification, accountability tracking, and response routing.

Sending and Receiving a Chunk of Information (Documents/Data)

There are frequent call center situations when the caller will need to receive or submit non-verbal information, such as documents, forms, reports, or product literature. (We are not talking about small data items, like an account balance number.) In the past, "snailmail," and, more recently, fax, have been the most practical means of document transmittal, and usually as a post-call session task initiated by the call center representative. Self-service IVR applications have also included options for information fax response where appropriate. Such responses could either be standard information common to all callers, or database information extracted into a customized report for transmission to a fax number.

Both forms of information are now more accessible to Web callers from their screen-based terminals, primarily for immediate receipt from a call center system via self-service Web applications, or manually delivered ("pushed") as a Web information page by a WSR action. Right there, Web callers get the benefit of immediate call center responsiveness that is better than previous technology offerings. However, what makes it even more practical is that when the WSR is actively connected during this process, the loop is closed for extremely efficient consultative interaction with the Web caller to bring closure to the caller contact. (Some folks like to call this "collaborative" interaction, but the customer relationship is based on assistance, not collaboration!)

How about the reverse, when it is the caller that needs to provide some type of digitized document (scanned, faxed, or emailed) or data files? Sending such information as attachments to Web mail from a self-service Web site application keeps the information together with the context of the Web caller's other information, making it much more efficient for response processing by call center staff. It helps eliminate the problems of delays or big mistakes that can result from the lack of supporting information associated with a Web caller's request, simply because they are transmitted separately. Here we see one of the major benefits of multimedia messaging, where messages and information objects are kept together during transmission and subsequent retrieval.

"Different Strokes For Different Folks!" - Not All Web Callers Are Configured Equally

With the traditional telephone caller, there was only one variable that had to be considered for providing self-service application functions, and that was touch-tone vs. rotary pulse dialing capability. Lately, Automatic Speech Recognition (ASR) capabilities have improved enough to enable voice input instead of just touch-tone. Once a voice connection was made to a CSR, there was absolutely no difference for the call center as to the type of telephone instrument the caller used.

Web callers, however, will be using different PC configurations that will indeed impact their options for call center assistance. In particular, the availability of a single telephone line vs. two phone lines or an ISDN connection, or low-speed modems, can make a big difference on the level of call center assistance that can be provided. Such variances will spell the difference between enabling a voice connection during a Web session or using "chat"- realtime text message exchange - instead. Voice conversation, however, is clearly much more time efficient and natural than chat, for both the WSR and the Web caller, but when the quality of internet telephony or the Web caller's PC configuration cannot support voice, chat communication may be a practical alternative.

The consequence of Web callers having different levels of communication capability means that the call center has to be prepared to serve them all. Call center technology providers are scrambling to develop "Web enabling" or "Webifying" capabilities, but the current products come in many flavors, with both subtle and significant differences between them that will impact both Web callers and WSRs. Such differences must be taken into consideration when planning call center capabilities to support Web-based activities.

The Three Web Caller Response Modes

If you read every call center vendor's so-called white papers, you will find them all plugging their ways for Web callers to reach call center support. What makes things confusing, however, are the various assumptions they make about what the Web callers' PC and phone line configurations are, and how they will "integrate" and "blend" Web call handling assignments with current call center ACD activities. Everyone has implementation strategies for accommodating Web caller support functions, which boil down (so far) to three basic forms of manual support and two forms of automated self-service. The former are extensions of traditional call center responsibilities for telephone access and text chat, as well as non-realtime email responses to Webmail. The self-service functions encompass both screen-based interactive (immediate) Web applications, as well as automated message responses to specific Webmail requests.

Insight: The call center response to Webmail message requests is interesting in that Web callers will not necessarily know whether their messages will be handled by a person or an automated application process. This could make a difference if that were known!

The need to be responsive to Web callers in immediate or deferred modes has attracted the attention of traditional call center operations to support such capabilities. However, doing so will have significant impact on the call handling staff, in terms of workstation software, skill requirements, blending of task assignments, and measures of performance.

Live call center responsiveness can logically include:

1. Immediate interactive response
    - Immediate callback telephone connection
    - Callthrough voice and data connection
    - Chat (Instant text message exchange)
2. Deferred or Scheduled Contact
    - Call center outbound telephone (voice) connection
    - Attempted connection next time Web caller is detected on the internet
3. Message response
    - Manual control text message response
    - Voice message attachment to email response

Depending on the various implementation strategies of the technology product providers, there will be different implications for both the Web callers and the WSRs.

Immediate Response

Technically speaking, callback strategies can be used by call centers for immediate or scheduled return calls from the call center. However, immediate response means providing live call center support (voice/video) at the time that the Web caller activates a request for assistance and while s/he is still connected to the internet and the Web site context. Maintaining the Internet connection is essential to obtaining the benefits of multimedia consultative interactions for both sales and customer service situations.

For an immediate voice connection, a couple of alternatives can be provided by available technology products. All start with the Web caller clicking on a call connection icon. The first approach, based on having separate telephone access to the Web caller, will trigger an outbound (callback) telephone call from the call center to the specified telephone number. However, this outbound call approach is the most expensive one for the call center, since it must pay for the calls without the benefit of toll-free or 900 access numbers. Straight telephone callbacks will certainly be costly in the global e-business marketplace, when international callers need assistance, unless some form of Internet telephony bypass is utilized. Finally, and most importantly, such outbound activity relies on the Web caller having a second telephone line.

A better methodology, callthrough, exploits Voice over Internet Telephony (VoIP) protocol, using H.323 standard "IP phone" software on the Web caller's PC. Client software that is efficiently downloaded (if necessary) as JAVA applets to the Web caller's browser, essentially enables IP telephony in conjunction with a gateway server. The most important benefit is that the Web caller doesn't require a second telephone line. The IP gateway server integrates with the call center's ACD, making it practical to direct such incoming voice connections to telephone call handling staff. This approach makes it perfect for accommodating callers coming in from anywhere on the World Wide Web. However, public internet traffic doesn't insure reliable voice quality, and other implementation strategies, such as switching the Web caller's PC connection from the internet to a dedicated "Intranet" access number, are available to insure a reliable and adequate path for voice conversations.

This gateway technology is also applicable to immediate callback situations as well, where new "Internet call waiting" services from local telcos can be exploited. Such technology enables the Web caller to accept an incoming telephone call via the local telco Central Office, while working on the Web and connected (on a single telephone line) to the Internet. However, bandwidth requirements for multimedia consultative connections, using today's common 28.8 KBPS modems, has left much to be desired for help desk technical support applications, as opposed to sales support interactions. The implementation of the T.120 document transmission standards will help improve this situation.

Interactive "Chat" and Web Site Visitors

Because Web callers are typically in desktop typing mode, it would seem practical to exploit chat text exchange for their interaction with call center staff. This might also be appropriate where the Web caller's PC is not configured to support Internet telephony and voice, such as a slow 14.4 KBPS modem. However, the relative inefficiency of half-duplex text exchange, compared to full duplex voice, makes this strategy highly questionable for maximizing call center staff productivity. Text exchange is a slower and less convenient means of communication expression for both parties than a voice conversation.

One application of text chat that is being promoted by chat technology and service providers is a Web site equivalent of outbound telemarketing activities. That is, based upon Web site visitor's actions, a call center representative can be assigned to initiate a text chat conversation with the Web caller. Monitoring Web site visitor actions is definitely a practical marketing strategy, but it doesn't necessarily mean that a text chat interchange is automatically called for. First, the detection of visitor activity should be done as a basic Web site application server function. Second, enabling the Web caller to initiate or allow any immediate, realtime contact is definitely part of the Web site self-service application, not a chat function. Finally, recognize that one of the perceived benefits that Web callers expect from the using the Web is the ability to be in full control and maintain maximum privacy over their activities. Any invasive response to their Web site experience will probably backfire, unless it is effectively qualified and presented properly.

Insight: Think about it! When would YOU rather use text chat to communicate with a stranger for e-business purposes, if there can be voice conversation alternatives? Furthermore, do you want highly skilled, expensive call center staff spending their time "cold messaging" with unqualified prospects? Preliminary estimates of comparisons of using text chat vs. a voice telephone conversation show that, on average, the text chat sessions are twice as long as the voice sessions.

Deferred Scheduled/Unscheduled Callbacks

The Web caller may want live assistance, but may not be able to get it during this Internet session. This will be particularly true if the caller is not configured for a voice connection, if there is going to be any significant wait (in queue), or there is a need to supply information for preparation on the part of call center staff. In such situations, rather than simply abandon the connection attempt, the Web caller can be offered the option to schedule a callback by telephone at a particular telephone number and time that would be most convenient. This information, typed into a Web form, can be submitted into a call center scheduling process, such as a traditional outbound dialer application, along with any context data (Web page information) that should be included in the "screen pop" for the call center representative.

Scheduled callbacks can not only be traditional telephone calls, but may also be "meet me" Internet appointments. Using software like NetMeeting and "Instant Messaging" connections to make contact on the internet, call center scheduling software will assign the outbound task to an available WSR, who will make contact with the customer. The connection will be an Internet data connection, with voice and data exchange ("white board") capabilities.

What will be even more practical is the unscheduled Internet callback. Rather than guess when the Web caller will be connected to the internet at a specified time, the "appointment" doesn't even have to be scheduled, but can be based simply on whenever the customer is next detected on the internet. Using internet service facilities like the "Buddy system" and "Instant Messaging", the call center's Web site application can be notified the next time the Web caller accesses the internet and it can reconfirm his/her need and availability for immediate call center assistance.

Telephone Callback in Response to Webmail

Before the internet promoted email for the masses, voice mail or IVR voice messaging was always a practical means for telephone callers to request a telephone callback response rather than wait in an ACD queue. That was in the days before public access, two-way messaging was available, and the telephone was the only rapid alternative to snail mail. Now, however, the response alternatives can be selected by the Web caller or the call center WSR. Either web callers, who send in a Web form or an email message, may receive a telephone call or an email message response, depending upon the nature of the request and the contact information provided. However, it should be noted that making a successful outbound telephone callback is still an iffy proposition, because, unless a specific callback appointment is scheduled, the callback attempt will end up with dialer retries. The messaging response alternative offers practical payoffs when dealing with Web callers for most applications.

Insight: It's not the speed of response that typically will require the telephone callback, but rather the complexity of the subject matter that requires a time-efficient discussion to gain understanding or negotiate agreement.

It's All Up to the Web Site Applications and the Web Caller!

The enterprise can inherently exercise control over the method and conditions of call center access by Web callers, just as publishing an 800 number for sales or customer service invites telephone caller access. However, it would be dumb to open a Web site for the marketplace and not have the proper means of supporting the expected Web callers' needs. Therefore, it is strategically necessary to thoroughly understand and prepare for the call center staffing issues that are associated with creating a Web site and supporting Web-based self-service applications. Depending on the needs and preferences of your enterprise callers, your plans for supporting Web callers will have to keep pace with their demands for the conveniences and payoffs of Web access.

Insight: Just remember, it's NOT the Web master who is running your business! Call center management must be responsible for all customer application interfaces that lead to live assistance, with both realtime voice connections and messaging responses. Unless there is are significant benefits for Web callers, in terms of convenience, saving time, or saving money, to access your Web site, they will still use the telephone, if they come at all!

Summary

There is no question that the Internet and the World Wide Web have created a revolution that will impact how customers and businesses will communicate. As one seasoned call center technology executive commented to me, e-business and having a Web site for sales and customer service is going to be "table stakes for staying in business." The telephone and voice communications will not disappear by any means, but person-to-person voice calls will become subservient to the greater flexibility of digital information exchange.

Web callers will be generating increasing amounts of call center traffic, and, rather than simply emulating telephone callers, they will also be exploiting both the informational self-service facilities of Web application servers, as well as the features and benefits of two-way messaging and multimedia information exchange. The challenge for today's call centers is to understand how this trend will affect their current operations and start planning for such changes. With Y2K issues hanging like a sword of Damocles over all computer-based systems, there will have to be a slow evolution from legacy technologies to the post-year 2000 future.

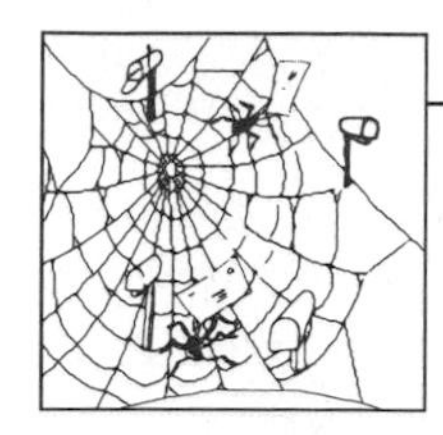

Chapter Four
The Call Center Meets "Webmail": Get Ready to Deal With Email Processing

The headlines and the statistics say it all! The world is moving swiftly into two-way digital messaging, which includes both personal email and enterprise "Webmail."

• Email has replaced traditional letter-writing and expensive long-distance telephone calls for many families, according to the Washington Post.

• Forrester Research projects that the number of email users will grow to 130 million in 2001 and to 170 million users in the year 2005. Such users will also be exploiting electronic commerce on the Web, and the use of "Webmail" for two-way customer messaging.

• Industry research indicates only 30 percent of Fortune 500 companies respond to questions directed to the Web site's Webmaster, which means a large number of customers aren't getting the response they expect.

• A recent study of a group of 258 companies with call centers that support email shows that those companies receive approximately 300,000 email messages per year, with 55 percent of the messages going straight to the call centers. The study indicated that these businesses expect to receive approximately 2 million messages per year by 2000. The study also showed that the current mean response time for handling email (the response being in the form of email or fax) was 17 hours, the mean cost of handling email was $17.85 per message, and that 54 percent of the cost is directly related to labor cost!

So, what does that all mean for the enterprise and its future call center operations? If it hasn't happened yet, it will happen soon. That is the day that your Web site manager will come to you with a problem. They want to find a cost-effective way to deal with customers who will make contact via the Web site, instead of the traditional telephone call. Why are they coming to you? Because you know all about communication traffic and managing the staffing to handle the influx responsively, and they want to know how your call center facilities can help manage the processing of email traffic from customers.

Of course, you will have to tell them that your legacy ACD switch system was never designed to control and manage this kind of message traffic, and new technology must be added to make email response processing cost-efficient. This is especially true when the Web site application can't always be "self-service" and completely automated.

"Web Mail" - Why Would Web Users Message an Enterprise?

The alternative to establishing a realtime connection with call center staff assistance, is through two-way messaging, except that the interaction is not person-to-person, but between enterprise workgroups and the Web caller. The workgroup can include not only people, but self-service applications as well, such as typically associated with "help desk" support.

There are many practical reasons why Web callers will choose to send a message instead of establishing a realtime voice connection with call center staff, including the following:

• They can't or don't want to wait for someone to assist them.

• They will use the context of web page information for the basis of their message inquiry or transaction.

• They need the convenience of 24x7 messaging access.

• They don't necessarily need to "discuss" anything about the information they want or the transaction requested.

• They may want to send information and data attachments first, before any interactive discussion is engaged in.

• They want a well-documented audit trail of the two-way messaging interchange.

• They want to eliminate the time, errors, and effort involved with transcribing detailed information during a live conversation

• They don't have a required second telephone line for an immediate live callback while still connected to the Web.

• They want to direct their communication to a very specific group or individual that may not be readily accessible realtime.

• It's just more convenient and cheaper than a live connection.

Messaging is a call center communication access option that Web callers can exploit to either request an informational message response or a scheduled telephone callback. Because Web callers will implicitly have their own personal email mailboxes, two-way messaging exchange is now a more viable alternative to live telephone connections. Not only is there public access to the Web caller's personal mailbox, but multimedia informational attachments, that might normally be mailed or faxed, can also be immediately emailed in response.

Implications for Call Center Activities

First, there is no question that enterprise "callers" will include a healthy percentage of "Web callers," who will connect through the enterprise Web site, for either immediate call center assistance, or a deferred response (telephone callback/message).

Second, processing email messages is not cost-efficient, unless a high degree of automated intelligence is used to minimize expensive labor.

Third, messaging responses to Web callers, while not necessarily as pressing as picking up a telephone call, must still be handled within a reasonable time frame. Replies should reasonably be expected to range from less than an hour to a "next day" response.

Fourth, both realtime voice connections and email messaging correspondence will remain logical alternatives for supporting customer contacts originating from the World Wide Web, and must therefore be managed as related responsibilities.

Finally, just as traditional call center staffing activities must be properly managed, from recruiting, training, scheduling, to performance monitoring, so too must message processing activities also be organized and managed in similar fashion. This is an area that call center people know a lot about, but must now apply differently to email messaging communications.

Call Centers and Message Processing

There has been a growing exploitation of (two-way) email messaging from Web site visitors, but it was not done within the context of traditional telephone call centers. Until recently, such email activity has been relatively unorganized and often supported outside of the traditional call center's operational domain, typically as part of the IT department's email or Web site responsibilities. However, call center managers and "application managers" are starting to realize that the objectives of Webmail processing are identical to those of traditional telephone call centers, and it is now important to understand the differences in the operational strategies involved in order to merge them with current call center facilities.

From a practical implementation perspective, it would be foolhardy to assume that providing message-based Web caller support is just the same as servicing regular telephone callers. For this reason, it will be wiser to approach Web message response services separately from the traditional telephone call handling tasks, so that such capabilities can evolve effectively at its own pace. While this means that the tasks that involve direct communication interfaces should kept separate initially, i.e., the "front ends", response resources, and procedures, basic call center infrastructures, however, such as knowledge bases, application processes, and caller database information, must be integrated from the get go.

Insight: Not only do you have learn what your Web callers need to supplement self-service application functions, but you also have to find out what the message traffic will be like, and how to plan for the WSR skills that will be needed for message handling.

Email vs. "Webmail"

New technology gets into trouble with potential users of the technology by misusing terminology, and it is important to differentiate traditional, person-to-person email messaging, from enterprise correspondence. Webmail is email that originates from a Web caller via an enterprise Web page, and is not addressed to any particular individual or personal mailbox. Although Webmail includes freeform text messages, it will be critical for efficient message response processing to structure such communications as much as possible with Web forms. Since Web callers will usually gain messaging access to the enterprise call center though a Web page, the structured format for messaging communications can be easily enforced. So, although freeform email messages will also be part of Webmail, particularly for follow-on messaging replies, structured Web forms will remain key to cost-efficient Webmail processing.

Where a single message response to a request is not adequate, "two-way messaging" (Reply) or a conversational call will be required to complete the interchange between the Web caller and the call center. Continuity with the Web caller's self-service activities, call conversations, and messaging transactions must be preserved, no matter who within the call center becomes involved with that caller, through the use of tracking numbers and identifiers (e.g., Web caller or company names), activity logs, and multimedia transaction folders.

Webmail Processing

Just as ANI, DNIS, and IVR facilities help identify and route callers to self-service or human-assisted call center applications, so too will a combination of message-oriented processes intelligently expedite the routing and disposition of incoming Webmail. These message-processing facilities can effectively work with your existing email network operations or with Internet email access.

1. Message Content Scanning and Routing - Automatic text analysis

provided by content mining software is useful for both structured and freeform message analysis, and can even be applied to language translation for global Web callers. Advanced artificial intelligence (AI) software using knowledge databases, and case-based reasoning, can analyze the full content of a message for disposition routing to fully automated response generators or staff-assisted processing. The use of such natural language processing can eliminate or minimize a very labor-intensive step and allows practical context-based priorities to be exercised before any message routing (e.g., skills based routing) is performed. Natural language context scanning will be particularly useful for follow-on freeform message interchanges, after an initial response to a more structured Web form.

2. Webmail Accountability - Proper tracking and accountability are the hallmarks of efficient AMD and Webmail management. Since messaging is traditionally fraught with sender insecurities and practical methods of determining accountability, Webmail management software must "intelligently" confirm message receipt to the sender, indicate a realistic response time, and provide a practical method for tracking the individual messages associated with a Web caller transaction for future reference.

In addition to scanning the information about the sender and the nature of the message, it is also important to check for other requests from the same originator, even when a tracking number is not included. This may uncover changes from a prior request, additional information for a prior request, or cancellation of the prior request. Repeated requests may indicate the urgency of the problem, and trigger special treatment, but should not blindly generate redundant responses.

When we use the term "message" for Web mail processing, it is not just the raw message itself, but the results of any automated pre-processing used to scan the contents and route the message to an appropriate WSR. Such pre-processing can include an abstraction of the message content, suggested "canned" response components from a knowledge base, and any other "intelligent" suggestions for efficiently completing the response process. In many cases, it may a final manual review for quality control of an automated response.

The AMD software will track message processing assignments, turnaround times, and the follow-on email activity that results from an initial Webmail response, etc., for purposes of evaluating the adequacy of response information content and WSR activity. The critical objective here is being able to continually expand the scope and quality of automated message processing.

3. Message Processing Assignment Queues - Traditional call centers are well known for their call assignment queues, where incoming callers wait for the "next available operator" with the prerequisite call handling skills. Incoming messages are treated in similar fashion, but with a different perspective of priority. Just as priorities escalate over time for telephone calls waiting in queue, so too will that apply to email messages being managed by an "Automatic Message Distribution" (AMD) software package that can interact with your existing email facilities. The AMD software will track assignments, turnaround times, and the follow-on email activity that results from an initial Webmail response, etc., for purposes of evaluating the adequacy of response information content and WSR activity. The critical objective here is being able to continually expand the scope and quality of automated message processing.

Several message assignment approaches can be taken. First, message "routing" distribution can be made to specific WSR mailboxes, and then it will be up to the individual WSR to process the message in a timely manner. A second alternative is to keep the messages in a shared, common queue displayed to a group of WSRs, letting them choose their next message assignment. (A "pull" approach.) For maximum management and control, however, the AMD software keeps all messages in queue, just like an ACD system, and delivers individual messages, one at a time, whenever a WSR indicates availability for a new task assignment (a "push" approach). Even after assignment, the messages are still controlled through the AMD software, and can be reassigned, if necessary, if not processed promptly. This insures that messages are not forgotten about, and their inherent priorities dynamically maintained. New, sophisticated AMD software products provide such centralized message management for applications that must process large message volumes "responsively."

4. Message Response Processing - Because the labor costs of dealing with reading and writing text messages can be so much higher than talking on the telephone, there is a driving need to automate as many aspects of Webmail message processing as possible. This will not only minimize staffing requirements, but will reduce the need for additional skills training.

The text-scanning technology employed to analyze email content for message routing is also used for selecting response information and generating the responses automatically, or presenting such information for manual review and approval by a WSR. Although Automatic Speech Recognition (ASR) technology has been applied in similar fashion for voice messages, it is a lot more efficient to deal with the textual information of email and Web forms. The "worst case" situation, however, is to have WSRs read Webmail (or listen to voice messages) and manually type a complete email response.

Fortunately, many frequent Webmail requests can be fully automated (like FAQ) or, while still requiring some customization by a WSR, can also be quickly disposed of with standard message templates and cut-and-paste knowledge base response information.

Realistically, however, Webmail processing requirements can encompass a spectrum of routing and response procedures, depending upon the Web caller's needs. These can include:

• Automated acknowledgement of message receipt ("mailbots"). These do not do anything with a Web caller's message, but send a canned response to every message received.

• Automated text content scanning and analysis, with fully automated email (or Fax) responses. The latter can still be "personalized", based upon database information. Given the many advantages of email (data attachments, privacy, security), fax responses to a standalone fax machine, or even to a mailbox, will be of questionable value.

• Automated text content scanning, with automated, skills-based assignment to an appropriate, available WSR.

• Automated routing to an available WSR, with manual content review and processing

• Response message templates, with "cut & paste" information from a knowledge database.

• Webmail can be quickly and painlessly transferred to more expert staff, complete with all necessary context information.

• Manual customized email response created by the WSR.

• Return telephone calls (callback), rather than an email response, based on complexity, need for discussion, and urgency of situation.

• "Instant Messaging" or "chat" connections can be activated if the Web Caller is logged on to the Internet; however, text messaging is not necessarily the most efficient way for call center staff to communicate with Web callers in real time.

• Where customized responses are always necessary, new "cross-media" messaging techniques can be applied, since it is faster and easier to create a voice message response (sent as an email attachment), which avoids the problems of spelling and typographical errors, and increases personalization of the response.

Mix and Match - Using Email for Immediate Call Center Assistance

One of the benefits of Internet-based email, is that URL access can be provided within the text message content. So, when the Web caller receives a call center email response, and live assistance is desired, immediate call center access can be initiated by the recipient from the email message, just like from a Web page. This approach also has the advantage of establishing the voice connection in conjunction with being on the Internet, and enabling multimedia consultative interaction.

Insight: Think about self-service IVR applications that let the telephone caller "escape" from the automated application to get immediate live voice assistance. This is also an "escape" from the messaging mode of Internet communications to immediate multimedia assistance.

The Importance of Contact Management and Message Logging Information

Customer contact information has always been important information available to telephone call handling staff, in order to maintain continuity of customer activity between contacts. The new wrinkle with email messaging is that it documents the exchange between the Web caller and the call center, not just for the call center, but also for the Web caller. Thus, the customer has a record of what was communicated, and it will be essential that the WSR have that same information at hand for any discussions that refer to past communication content.

Because Web callers can easily switch between email communications and realtime voice connections, it will be necessary to track all such activities in order to stay in synch with the most current status of the Web caller's communications. For example, what happens when Webmail is sent, but when the Web caller finds no response in their mailbox, they immediately get on the telephone, or go to the Web site and request live assistance, or send another Webmail missive. Clearly, the operational software has to recognize this set of circumstances, apply suitable routing logic, and provide the WSR with appropriate information to handle the situation effectively and efficiently.

WSR Message Processing Skill Requirements

As indicated earlier, Web caller text message processing must be as automated as possible in order to be cost efficient timewise, as compared to voice call handling. This is particularly necessary to support the following kinds of skills necessary for message handling:

- Email message management - Sending, retrieving, addressing, routing/ forwarding, replying, archiving, data attachment handling, etc.
- Written language skills - Reading and screening Web caller text content, grammatically correct writing, selective use of technical terminology, properly employing customer-sensitive verbiage, etc. For global E-business, multilingual language skills, as well as being able to understand poor English text, will be an important consideration.
- Typing proficiency - Speed and error-free typing to minimize the time to create a text message response. Accurate name spelling and data entry will be extremely critical, as usual. This includes accuracy in procedural keyboard entries for email messaging and data file management.
- PC usage literacy
- Internet/Web site navigation skills
- Web caller applications procedures and information
- Understanding Web caller perspectives, mindsets, and psychology

Text message processing is still going to be more time consuming than voice, even with the use of skilled personnel, and, because such personnel will be now be more expensive to recruit and retain, their time has to be used much more selectively on an exception basis. That is why the real solution is to be much more organized and automated for email response processing.

We should also make note of the fact that the skill requirements mentioned above are also necessary, and perhaps even more demanding, to support interactive "chat" sessions with a Web caller. Because the WSR must rely primarily on their own writing and typing skills in real time, there is clearly a higher performance requirement, less manageability, and more questionable benefits for WSR staffing, when compared with a voice interchange.

Keeping the Voice of the WSR in Email

The new text messaging processing and management software described earlier will be extremely useful in providing and managing automated solutions for the messaging activities of Web callers. In particular, standard problems and questions can be dealt with packaged responses. It will be the "exceptions" that require the personal attention of a WSR, whether because of the level of authority or special expertise that is required for a response, the need to "personalize" the response, or because the Web caller's needs are simply something new that doesn't yet have a "canned" response. Although the numbers may vary from application to application, the need for WSR involvement with email responses may typically be reduced to below 20% of the total email traffic. ????

But, what about that remaining 20% or those applications where the human touch is always needed? The predicted increase in email traffic will be so big, and the demand for highly qualified "gold-collar" workers so great, that any strategy for increasing their productivity and making their tasks easier should be vigorously pursued by call center management. One simple alternative, the continued use of voice for personalized messaging responses, should help provide those kinds of benefits for WSRs.

Insight: It's the message response customized content that can be time-consuming because of the writing, spelling, and typing skills required for email response efficiency. The voice skills of telephone call handling, however, are easier to find and train, and are ultimately more time-efficient to use operationally. From the recipient's perspective, voice messages will be acceptable and have more "personality" than cold text.

By the time this is published, I wouldn't be surprised to see call center product announcements that enable call center staff to respond to email messages by simply recording their responses (as .WAV attachments to the response message). For those Web callers who need it, "player" plug-ins can accompany the message to the recipient's mailbox. So, what's the deal? Simple! The WSR can review the Web callers needs, record the voice response more quickly than by typing it, AND, if necessary, still attach data files, documents, etc. to the response message. The bottom line is that we have made it easier for WSRs to communicate with natural voice, including multilingual situations, rather than having them get a college degree just for the writing skills that would be necessary.

What about the audit trail provided by text messages? If this is indeed needed, the "new and improved" Automatic Speech Recognition (ASR) software can be put to good use here, especially since it doesn't even have to be so speaker-independent. (After all, call centers will be using regular employees to support their critical operations.)

The IVR Connection: Web Forms vs. Voice Input

There is a similarity between application-structured Web forms for capturing caller request information and the use of Interactive Voice Response (IVR) voice scripts for capturing similar information as recorded voice or DTMF data for call center applications. The latter may be numeric data or menu item selections.

The limitations of the voice interface include:

• Information selection limited to short lists, in order to simplify user interface

• Information collection time-limited to insure acceptable user interface

• All information input must generally be included in pre-programmed scripts

• Freeform information is recorded voice input, which must then be manually transcribed

Although speech technologies like Text-to-Speech (TTS) and Automatic Speech Recognition (ASR) can alleviate some of the above constraints for traditional telephone callers, the Web Caller will already be in screen mode, rather than in voice mode. Interactive Web forms provide advantages over freeform email messages by enforcing required information input and providing interactive error checking of selected entry fields. Input guidance, help information, and comprehensive menu displays all make screen-based input much more time-efficient for the Web caller.

The screen user interface lends itself more easily to self-service applications, Interactive Voice Response (IVR) technology has expanded to accommodate Web callers, by also becoming the development tool for creating and processing the user interface for Web forms that integrate with traditional call center functions. Since there needs to be consistency in content given to telephone callers and Web callers, a common call center application can now be used to drive both the voice and screen caller interfaces in a similar, but not identical manner.

Where Does "Unified Messaging" Fit Into the Call Center?

It's unfortunate that the fast pace of digital information and communications access has created so much confusion. The concept of "unified messaging" pertains primarily to the need for consolidation of personal email, voicemail, and fax message management, either via a common message store (mailbox), integration and message "synchronization" between message servers, or by coordination through the user interface client software.

However, "call center messages" are not personal, cannot be controlled entirely by the individual WSRs, and may well have different priorities and routing rules for processing, depending upon the form of access involved. For example, an incoming fax message, which is a form of "one-way" messaging, will not be as demanding as a Web caller waiting for a voice connection, but may be considered more pressing than an email message. It will be tne routing "queues", rather than a unified mailbox, where message management logic will be applied for voice/fax/email messages that come to the call center separately. However, because Web callers will receive and may send multimedia email messages, such messages must be managed as email (with attachments) for call center routing, retrieval, and archiving. Although the IS department would view Web voice calls and message management as simplistic "groupware" or "workflow" applications, realistic call center management will want realtime facilities to monitor and assign WSR resources for both message processing and voice call handling traffic.

Insight: The telecommunication industry is still struggling with the definition of "unified messaging," in terms of what it means to the end user vs. how it can be implemented. The call center's focus will be on multimedia message exchange at the user desktop, which email technology already supports.

Managing Webmail Processing

Typical call center management responsibilities will apply to staffing for Web mail processing. These include recruiting, training, work force scheduling, and performance monitoring. In addition, management must also take responsibility for specifying the rules that will be applied to message routing and disposition, based upon content analysis and elapsed time priorities.

Call center management has already developed most of the tools necessary for managing voice call handling, and some of these may be adapted to email message processing. However, new management performance criteria and metrics will be involved, and software facilities are coming from the call center industry to support such added requirements.

Web Application and WSR Performance Monitoring and Evaluation - Just as call centers must monitor and evaluate the performance of their traditional telephone call-handling activities and "applications", i.e., IVR call flows/information content, CSR scripts, and individual CSR performance, so, too, must there be auditing facilities for regularly evaluating the activities associated with Web callers. From the Web caller's perspective, there may be Web page applications (self-service), and both voice connections and email correspondence to be evaluated. In other words, a total transaction has to be taken into consideration, in order to properly evaluate the impact on a Web caller. This will require collecting data from each of these activity areas into a unified picture of all events in a completed Web caller transaction.

Insight: Just when we thought we were getting "end-to-end" telephone call activity consolidated with better, "open" call center technology, now we have new functions that have to be included in the call center activity stew.

Call monitoring technology products allow the selective recording and collection of voice conversations, along with associated desktop screen actions performed by call handling staff. Such capability should take care of real time Web caller connections handled by WSRs. What must be added in are the Web callers activities on the Web page, which is already being collected for proper call handling purposes, and the activity data associated with email response processing.

For the moment, these will not be quickly "integrated" into any consolidated reports, just as the industry had to wait for IVR activity to be tied into live call handling reports. However, with "open systems" and network access being de rigeur for the future of all call centers, it is only going to be a matter of time.

Adding Webmail Processing to Your Call Center

The technology to manage Webmail processing is becoming available from various sources, including traditional call center and ACD system providers. Two approaches are being touted; one that separates Webmail message processing from realtime call processing for Web callers, the other attempts to "blend" message processing assignments with live calls (normal telephone call traffic and/or Web Caller outbound "callbacks" or inbound "callthroughs").

There are several practical reasons to go slowly in merging voice call processing with Web mail processing in the call center.

1. There will be much to learn about Webmail processing, traffic, staffing, and skill requirements, for your applications. Until they are fully understood, the traditional call center operations should not be exposed to any perturbations from these new activities.

2. The critical application-based automated processes are going to be very difficult to design and implement, even with the best software tools that are starting to become available. Its not a matter of programming, but rather an issue of heavy-duty analysis of Web caller needs, both past and future, and insightful understanding of effective solutions. This responsibility belongs in the bailiwick of the "application manager," who must fully understand where Web callers are coming from. If you have never

done this kind of thing before, you better get help from those who have!

3. Realtime connections ("callback," "callthrough") should also be cautiously implemented in order to learn how best to handle the "immediate" needs of your Web callers. Because there will be more of an impact from this activity on staffing and system resources, depending upon the applications involved, i.e., sales, customer service, Webmail processing may be a more practical starting point.

4. Even though these different caller interfaces should be implemented individually, the "back end" information databases must be consolidated and shared in a consistent way. Thus, all customer information, activity logs, and even past messages, must be accessible and strategically utilized for all self-service applications and CSR-assisted tasks.

Once all Web Caller support activities have been successfully implemented as separate operational functions, it will be appropriate to evaluate the ways in which they can be consolidated and "blended" from the perspective of shared staffing and relative assignment/routing priorities. Even though call center personnel can handle different types of task assignments separately, it may be more complicated and ineffective to intermix them sequentially or multitask them.

Initial Email Traffic Statistics

Web eBusiness is still in the transition stage and evolving towards a more steady state of multimedia communication activity. This will only happen after the technology shift has been completed, both from the callers' side as well as at the call center end. The following early data reported cannot be considered conclusive, but may be interesting to track for the future:

• There does not appear to be a correlation at this time between the amount of incoming voice traffic and email traffic. This may mean simply that email traffic is a new source of additional communication contact by Web callers, that has no significant impact (yet) on traditional telephone traffic. Some call center operations report 20-25% of all incoming traffic is email. However, for pure web-based activities, the call center statistics shift in favor of email; one such operation reports 88% email, 8% chat, and 4% live calls. This distribution may reflect the requirement for the Web caller to have two phone lines in order to effect a voice contact, while remaining connected to the Internet.

• Although the numbers will vary with individual applications, typical fully automated email responses constitute 20% of the responses, while 80% involve human effort. Simple queries for information retrieval, not requiring judgement or decision-making, are the primary targets for such fully automated responses. Of course, some of this kind of information may also be provided directly on-line by an enterprise web server application.

• Email processing times range from 30 seconds for inquiries requiring nothing more than a "canned" response to 10 minutes for messages requiring research of multiple "back office" databases for customer information, etc. At this stage of initial implementation, there is less concern with processing time than meeting response time service levels.

• The need for a "follow-up" email exchange after the initial email response is not known, and most operations attempt to avoid such situations. One industry provider estimates this happening perhaps with about 10% of the responses.

• Typical email response times today (using email processing technologies and WSR staff) aim for a "next-day" objective, but actually range anywhere between one hour and two days. However, a same day response objective is a current target for agent-assisted responses, and fully automated responses being immediate.

• Various measures of email processing performance are being utilized, although often in relatively simple ways. Such measures include overall service level attainment, number of messages processed, average time per message, quality of response message content, etc. The technology for making this task easier is still evolving.

Summary

The World Wide Web and Internet email will require that the power of two-way messaging be added to the repertoire of call center. Using two-way messaging technology expands upon the sophistication traditionally found in "help desk" environments, and makes such technology useful for almost any call center application. This will ultimately prove to be a major benefit to the call center's most important operational problem, staffing and labor costs. It will offload call center operations away from the staffing pressures of the telephone traffic "busy hour", towards more manageable and cost-effective support activities.

However, because of the increased skill requirements for processing text messaging, new technology tools and the use of email-based voice messaging will be an absolute necessity. Such capabilities can be added to supplement traditional telephone call centers, with either legacy ACD systems or the newer, network-based "all-in-one" communication servers.

Chapter Five
Technical Intimacy and the Art of Customer Loyalty

Introduction

The "call center" or "contact center" or "transaction center" is the point of engagement where your company's network meets the great wide-open public network. It is where your enterprise's nervous system interfaces with your customer's touch-point. These touch-points and the customers that use them are increasingly exploiting the power and convenience of digital communications for information and personal contact; supporting the blistering schedule of a world that is open for business 24 hours a day. The call center will be the focal point - the front lines of frustration where customers are defined and dealt with.

The fundamental premise throughout this chapter is that the only way to adapt to the accelerating pace of change is by overcoming the natural resistance to such change and by discarding obsolete concepts and processes. Provocative and practical operational insights on the strategic role of new call center technologies are introduced that will undoubtedly generate controversy in the industry.

As this chapter was going to print, I discovered a report developed and written by PW Coopers consulting firm. The firm studied a number of North American companies and came to the conclusion that all companies could be stratified into three groups, or "stages of evolution."

The Stages of Customer Evolution

The PW Coopers "IDEAS" study found three stages in their model of the "Stages of Customer Evolution:"

Stage One - Customer Acquisition 35%. 35 percent of North American Companies are in Stage One. Stage One companies can be characterized as "hunters." They are focused entirely on customer acquisition; hunters swoop in on the opportunity kill with little or no thought to their next meal, margin or customer.

Stage Two - Customer Relationship Management 59%. Stage Two companies are considered "ranchers." Ranchers cultivate their customer margins for margin slaughter later. Many more companies are in this category and do not want to realize it.

Stage Three - Customer Asset Management 6%. Stage Three enterprises cultivate customers using sustained yield management practices. This paper outlines some of the defining characteristics of an enterprise that is practicing true sustained yield management.

While this PW Coopers definition of the stages of customer evolution may be simplistic (and even if it is off by even a few percentage points), it is an eye-popping effective model in its demonstration of how 50 percent of companies believe that they are actually in the top 6 percent class of companies!

These three stages also reflect the evolutionary shift from measuring day-to-day transactional efficiencies to gaining lifetime customer intimacies. The difference between these strategies is why a Home Depot can know more about you than your dentist does. Interestingly, these 6% companies that are implementing progressive customer intimacy strategies more effectively than others are also doing much more outsourcing of their non-core competencies.

Expectations

Call centers are, without question, at the heart of today's consumer-focused enterprise. The three or five minutes that a customer is engaged with your enterprise represents a "moment of truth;" the moment that, if your customer's needs are not met expeditiously and professionally, your company stands to sacrifice quality in the customer relationship.

Rising expectations from callers and customers are being created by falling interaction and communication costs. The "bar" of expectations is being raised through the emerging complexity of customer engagements (the internet has guaranteed that customers are more "knowledge-ified" than ever) and their increasing expectation for tiers of service - an expectation developed by airline frequent flyer programs, for example.

These expectations of service are a direct result of the power of the computer because we now can keep track of every transaction, conversation and detail about our customers. This transactional and dialog detail enables an enterprise to execute strategies based on the concepts of mass customization or relationship marketing. Establishing an interactive experience that is characterized by an obsession for reliability, speed and intimacy is the most significant way to effect and forge the relationships that will characterize the successful enterprise of the next century.These expectations of speed, reliability and intimacy are also increasingly becoming the benchmark that your customers are using to measure the quality of the interaction with your enterprise.

Make It Consistent - No Time For Downtime

It is amazing how many call center managers and directors "check" their real-world experiences at the door of their enterprises when they come to work. Most call center managers may recognize themselves for just that - managers, but as consumers with their own hectic, confused daily routines, they tend to forget that they too live in a world steeped in their own customer interactions.

Although this is a small space for a long subject, many managers miss the point that caller expectations are not set by your industry-specific competitors; they are set by other call centers. What we often fail to realize is that we now compete with better service from places we never suspected. A common mistake made when benchmarking call center services is peer comparison; if you are a bank, you tend to compare yourself to other banks; if you are a utility company you tend to compare yourself to other utilities; insurance, retail, and help desks are all apt to compare themselves to their respective rivals. Why? Your customers don't. Your competition is no longer from the usual competitors. You now compete with Citibank, Disney, Fox, Brand X, Microsoft, USAA, the electric company, my bank and VISA, who comprise some of your "expectation" competitors, not your same-industry rivals. Consistency is not measured from transaction to transaction, it is now measured from company to company.

The relationship line between the customer and the company is moving, blurring and changing in different directions at once. Increasingly, you must look further and further outside your own business segments to discover where expectations about service are being created. When Home Depot knows more about your customer than you do, it is a sign that a fundamental shift in the power of expectations is happening. A key consumer shift caused by technology is that every customer now has the ability to survey the market place and make optimal purchases - a knowledge-ified power over the market.

Customers and enterprises alike have developed an appreciation of the potential of technology - we know when we want to use it (account balance retrieval, for example) and when it is poorly designed (long IVR menus). Expectations about www.com are now the same as those of an 800 number. The expectation problem emerges when managers refuse change or ignore what happens when technology is misused. Missed connections (ringing, then a busy signal), broken transfers and improperly handled customer requirements can put reliability expectations seriously at risk and can certainly cost the enterprise much more to repair, if even possible, than having done it right the first time.

Make It Quick - A Rebate In Freed Up Time

Time has become the most precious resource in the customer's life. "The entire economy is riddled with time-wasting routines and regimes that squander much of the time of the average customer. Suffice it to say that the concept of the customer's average life-span as a crucially scarce resource, indeed the most precious resource of the information economy, has not penetrated to many enterprises." (Gilder, Forbes ASAP 4/6/98, p.76). Higher clock speeds are giving consumers a rebate in freed-up time, and the need for speed is an expectation that comes from the time-wasting routines we have in our daily lives. So much so that consumers today will pay a premium for speed whenever they can buy it in airline e-tickets, microwave ovens, computers with modems, on-line banking, wireless toll collectors, and credit card swipes at gas stations. The customer who exploits the speed of new consumer technology is no longer willing to put up with slow lines or queues, filling out complicated forms, or telling tele-marketers whether they have had a nice day or not. As Gilder says, "Are you a life-span extender or a life-spam vendor?"

Customers are looking for business relationships that give them a rebate in time. At least this is my justification to my wife paying for lawn service, when I really just hate mowing the damn thing. Too many enterprises today waste our time; they make us drive to the bank, grocery store, bookstore, post office, telephone company, software vendor and schools. All make us line up in a queues, literally.

The fact is, if you keep your customers waiting, you won't keep your customers. However, be leery of the misnomer of "speed." If you do it fast but you don't do it right, you have sped up nothing. Focusing on transactional efficiencies over customer intimacies is the slippery slope of substituting activity for achievement. If you know much about your customers but are slow getting it, then it is useless to you. The slowest thing customers hate is not being able to reach you at all.

Make it Accurate

Customers have become so accustomed to the technology that they are now impatient if an enterprise does not know everything about their histories and transactions as soon as their agent answers the phone. Consumers do not want to waste time communicating unnecessarily - so you had better have something meaningful to say. The convenience of doing so is reflected in the service you provide. Done correctly and remembering the dialog you had with a customer the last time you talked, you can create a "barrier of inconvenience" - the reason for your customer to never want to deal with your competitor again.

As Harry Newton (we explain him later) wrote before he exited the computer telephony industry stage left; "So stop asking me, your loyal customer, who I am or how I plan to pay. You already know this; just connect your phone line to your database and you will get me off of the line faster and I will call back sooner." Make it consistent, make it quick and make it accurate. Never require the customer to tell you the same thing twice. Customers do not want noise, they want the news.

The Complaint

Customer satisfaction is at an all time low.

Harvard Business Review contends, in a recent series of articles and studies, that customer satisfaction is at an all time low. "Relationships between companies and customers are troubled at best." People are getting fed up. Complaints, boycotts and other expressions on consumer discontent are on the rise. Cell phone hell; voice mail jail; call center limbo are easily recognized examples. Customers believe that companies are disorganized, unresponsive and uncaring - even if they make reasonably good products. This is a reality few executives are willing to admit.

The slowest moving target of all is a competitor's unhappy customer and many smart companies are turning to call centers, web sites, and interactive software applications to bond with their competitors slow-moving, unhappy customer. We are increasingly hearing about confusing, insensitive and manipulative companies which trap and victimize their customers through mazes of technology. Which works, because from a consumer perspective, consumers are even less sophisticated than ever. Cheap PCs create more illiterate users, unfortunately.

Customers Want to Complain

My father, a respected and retired military person himself, verified that the generals and admirals at the Pentagon may just be right when speculating that the absence of a global enemy has made Americans turn on themselves. Because the economy is good, one of our focuses is on service. Customer want to complain and you should make it easy for them. In the one-to-one model if you get this wrong, you are shut out of a dialogue with customers that could teach you about how their tastes and preferences are evolving over time.

The Most Damaging Competitive Tool is the Service or Product Forum

Want to know what the most damaging competitive tool going nowadays is? It's the product or service forum: sometimes it's ad hoc and sometimes it could even be created and threaded along by your rival.

Customers want to complain and the internet-based service or product forum is one of the easiest ways to listen to them.

Customer service has always been a double-edged sword, and technology makes this even more pointed. Jeff Bezos, president of Amazon.com said, "If you create an unhappy customer on the Internet, they don't tell six friends, they tell 6,000." News travels fast these days and knowing that they are not alone with their problems empowers customers even more. A single grudge page can hurt a company in ways that far exceed the value of what is being done.

While electronic communities may contain a good amount of flaming and complaining, they can also contain powerful and articulate insights and observations that would otherwise be impossible to come by in the course of traditional customer interactions.

Mining the electronic communities for terrific and significant ideas could benefit your enterprise and customers greatly. An enterprise's marketing strategy, indeed its mission, should be achieving a "feedback loop" where customers perceive that they are known, remembered, listened to and catered to. Customer satisfaction can be garnered by sending a signal that your enterprise is in a constant state of improvement - a signal that you are busy doing what you don't know how to do in order to learn how to do it.

Insistence Isn't Resistance

A minor but annoying characteristic of technology and product managers is to view negative feedback from customers as a temporary resistance to change. Don't fall into this trap because customer trust is hard to win and even easier to lose. Customers will cope and tolerate companies soliciting them for information, if management really acts on it. A real-world example of this phenomena is with Microsoft's Windows 98. How many of us really want Windows 98?

Seeking The Pareto Complainer

The whole genre of one-to-one marketing was created by Dr. Martha Rogers and Don Peppers on the philosophical ideas of a 19th century Italian economist and sociologist named Vilfredo Pareto (1848-1923). Mr. Pareto discovered the 80/20 rule that essentially says that 20 percent of your customers account for 80 percent of your revenues or profits. Every industry today has "Pareto" trends. The "Pareto Customer" is the customer in a significantly more valuable class of customer than most others. The "Pareto Complainer," is the customer who generates a lot of value but also does not have a problem telling you what you could be doing to do what you do better.

British Airways knows that 35 percent of their customers account for more than 60 percent of their sales. 20 to 25 percent of cellular customers account for 60 to 80 percent of revenues; retailers find that 20 to 35 percent of their customers generate 70 to 80 percent of their profits. A recent study by Mercer Management Consulting showed that banks do not make any money on 40 to 50 percent of their retail customers. Credit card companies, clothing retailers and stock brokerages all have distinct segments of Pareto and unprofitable customers. Call AT&T with a question about your long distance service and the company will route you to any one of several different call centers. Based on your usage history, if your bills are high you will get what AT&T calls "hot-towel" service - a live, human operator. Spend less than three dollars per month and you get the automated routine.

Since building relationships is not a cost-free proposition, customers with the highest potential to perform should receive the most attention. Marketing becomes important to not bringing in the loser in the first place. One time customers or those that are identified as churners represent losing investments. Companies like AT&T have gotten much smarter about separating the "cream", the Pareto customers from those they don't want.

Competitive companies are extending quality customer engagements into life-long customer relationships, because it does not take much to realize that a satisfied customer is more likely to become a repeat customer. We in the call center industry have been inundated over the past several years with the one-to-one customer intimacy marketing message. It is a message steeped in the observation of how expensive it is today to find new customers and how much more efficient it is to keep existing customers. Propagated in the mainstream business strategy circles as the Loyalty Effect and the Peppers and Rogers phenomena, one-to-one marketing includes the intimate tracking of customer histories, mining databases and proactively anticipating customer needs.

A Few Percentage Points in Retention Equals $M's in Revenues

The change of just a few percentage points of customer retention can equal millions or more of dollars to the enterprise. Customer acquisition is 5 times more costly than retention and the average company loses 25 percent of its customers per year. The more products that you have to hook your customer with, the more loyal they will stay. A GTE study found that customer churn for customers with one product was 50 percent per year. For those customers that had two products churn dropped to 25 percent per year. Customers with three or more products had dramatic churn rates of below 5 percent! There are plenty of books out there that have all of the Harvard Business School level math, but the math has been done and acquisition vs. retention is a very proven and stable business equation from revenue and profitability standpoint.

The biggest challenge of one-to-one is putting the squeeze on losers and not offending your Pareto customers. Where marketing becomes important is in avoiding bad customers in the first place, then trying to find ways to get rid of them. So, when companies focus more on creating deeper relationships with existing customers than on attracting new ones, marketing will become a less visible engagement. It will develop into a private, intimate affair between the customer and the enterprises that they buy from.

The Pareto Customer - One-to-One Too Much

Enterprise Relationship Management (ERM) projects, an enterprise's MBA term for one-to-one marketing - are hard to do, take a long time to implement, are very expensive, require strong executive support and require considerable downstream administration. The construct of customer intimacy and one-to-one marketing is a valid idea. However, at the end of a seminar or conference, the managers of call centers will go back to deal with the operational issues of transactional efficiencies and one-to-one marketing strategies will have little influence on the decisions they make next week. It is neat stuff, but conference attendees don't do, instigate or manage the type of ERM projects that one-to-one requires to be successful.

Not to be ignored, though, in this 1:1 model is the mistake of underestimating that a customer may be a valuable source of new ideas, as a reference, or as someone who spreads a positive message about the enterprise through word of mouth. Communication Week won't give me, a technology writer with thousands of readers, a free subscription to their magazine because I am not the CEO, CIO or buyer of technology in a company with more than 500 employees and $30 million in technology purchases.

Proactive dialing is one of the underestimated technologies of customer intimacy. The DoubleTree Hotel in Austin uses a predictive dialer to call me regularly and ask me when I will be back in town and if there was anything they could do for me. Hotels are one of the better indicator industries to observe the pace of relationship marketing because hotels are most interested in determining their "acquisition-versus-retention" expenditures. They want to know exactly how much they should spend to grow their customer base.

The Theory of Intimacy

Loyalty is different than no choice; utilities and phone companies are examples. The theory of intimacy says that switching to another provider of anything should get harder and harder for the customer as they are increasingly offered service tailored to their specific needs. After building a relationship with a customer and learning all there is to know about them, there is little chance that they will leave for a competitor even if offered a better price. The enterprise must weave a web that is hard to break and as services become individualized, lines of comparison will become blurred making it harder for the customer to compare apples to apples.

The new span and power of technology makes it possible for every enterprise to remember each customer as readily as the customer can remember an enterprise – like the proverbial old corner grocer could. The logic is that by having as much knowledge about its customers as possible, an enterprise can more effectively serve and interact with its customers, improving customer retention and enterprise profitability.

Loyal Customers are the Most Unsatisfied

"It is startling how wrong we've been about what it takes to cultivate intimate relationships with customers." says the Harvard Business Review in January, 1998. A study by the University of North Carolina at Greensboro found that your enterprise's best customers are also more difficult to please. Roughly five percent of a large US bank's customer service calls are from its best, most profitable customers and, while these customers typically receive special services, they abandon their calls at more than twice the rate of the average customer. The most loyal customers were actually the least satisfied. These most loyal customers may expect more, perhaps because of better knowledge of your enterprise's products and services or because of what they have experienced in the past. The point is clear: when talking to your customers, it certainly helps to know how loyal they are . It will help you interpret their messages.

Relationship marketing is powerful in theory but troubled in practice. Misguided manipulation of sensitive information has sparked a consumers backlash that endangers the very credibility and reputation of relationship marketing. Enterprises are consistently underestimating the new ethical challenges the use of data warehousing introduces in the enterprise/customer relationship.

Data-Mining: Information is Now More Valuable Than the Real Estate or Technology It Sits In

The value of the information surrounding a transaction may now be worth more than the goods or services being transacted. In 1996 Chase Bank discovered that the value of the technology in Manhattan was worth more than the real estate that it sat on. Not too long afterwards, they discovered that value of their information may be worth more that the technology it resides in. It is no longer a battle for the corner office, it is a battle for the corner server - the one with all of the data.

Data mining helps an enterprise understand who its customers are, what they are doing or not doing, and if the enterprise is delivering against the customers' expectations. Data mining is used to produce high- quality business intelligence, such as association, classification and regression analysis to determine these customers that have a higher propensity to spend money with you. It was the mining of data that did overturn the conventional wisdom that the customers of banks who had multiple accounts were the bank's best customers; in fact, they are actually unprofitable. The enterprise that formalizes the way it sorts, analyzes and applies its data will have the competitive advantage. In the end, he who has the biggest database and, knows how to use it effectively, shall win.

Information Exhaust, Thousands of Attributes

Companies that are looking to turn data into new products have to begin by sifting through the enormous quantities of "information exhaust." From the vast numbers and combinations, very fine opportunity segments can be identified. One consumer credit card company has almost 1,000 attributes, or data-points per customer. One insurer tracks 150 variables with 40 to 50 parameters on some 6 million customers. A single mortgage application can have some 300 data-points per application. The software of data mining can evaluate millions of records with hundreds of variables. Wal-Mart leads the trend for deeper thoughts on using raw customer data to identify and exploit very fine opportunity segments.

Wal-Mart analyzes each of its 90 million transactions per week and manipulates a year's worth of cash-register transactions, right down to each shopping cart to see how purchases of different items are related. The computers Wal-Mart uses are able to analyze up to 700 million possible, price, store and item combinations. Wal-Mart spends more money on their on information technology, more than $500 million, than the GNP of many countries. They have recently tripled the size of their data warehouse to 24 terabytes and now have the largest commercial database in the world, second only in size to the US government.

Wal-Mart knows that the customer who buys Barbie - which it sells one of every three seconds, has a 60% likelihood of buying one of three types of candy bars. Wal-Mart is looking to identify products that sell well together so the company can display them next to each other. If someone wants to buy toothpaste, it is on the shelves. But, if they don't want to buy it all of the time, Wal-Mart does not want to have any more on the shelves than needed because it costs money for the inventory. Which leads me to wonder about the seasonal variations in the use of toothpaste and mouthwash. Which seasons and what parts of the country is what I would like to know!?

Big Guys, Little Guys; Technology Levels the Playing Field

What makes data warehousing uniquely relevant to call centers is this convergence (rather more like an intersection) of low cost storage, very fast processors and parallel processing applications that make it possible for data mining to ask new questions about more data. Moore's Law (Moore's law is about the doubling of processing horsepower every eighteen months) allows the little guys to increasingly do what only the big guys, like a Wal-Mart, could do.

Now it is possible to keep in a database the kind of information the old corner grocer used to keep in his head such as recognizing and doing favors for his customers. Although I have not been alive long enough to know what a corner grocer was, apparently we have gone a full circle and found ourselves now doing, because technology now allows us to keep track of customer dialogues and interactions, what the old corner grocer did mentally. Enterprises today have tens of thousands of customers, yet for the most part, these companies do not have a clue as to what makes their customers tick.

Panty Hose and Jelly Donuts

L'eggs pantyhose reviewed some of the customer comments left in their forum section of their web page and they discovered the popularity of pantyhose among men. They were surprised to learn that a select group of men like the warmth and comfort of these traditionally female-only products. Mining their information exhaust they discovered three distinct opportunity segments; law enforcement officers, hunters in the north east and serious outdoor athletes. Maybe L'eggs should consider a camouflaged king-sized brand. Until relatively recently, it would have been difficult for enterprises that practice customer intimacy to have discovered this new market segment.

Take the GM television commercial for Saturn cars where the jelly donut junkie has a history of complaining to the dealership about all sorts of imagined car ailments. The Saturn dealer, scanning his customer database discovers that what this customer really wants is the jelly donuts in the customer lobby. This brings up an interesting question about GM's use of customer data, because I'm not sure that customers would want GM mining their extensive data base for obnoxious jelly donut scammers.

The question that Michael Schrage brought up in a recent column was what happens when your records become annotated by customer service agents with comments like obnoxious, complainer, whiney or manipulative? This is the antithetical treatment that we are giving when we empower the agent taking the call with a subjective power like this. How does a company use databases to defend against abuse of service guarantees, like jelly donuts?

It is through the call center or point of customer contact that this friction will occur. What if it is the agent that is having a bad day?

Privacy: The Key Consumer-Protection Issue of the Information Age

Privacy is the key consumer-protection issue of the information age. Privacy is more important an issue than many think in the relationship marketing revolution. Privacy and intimacy are closely related and the problem is that businesses today don't tell consumers what they are doing with personal information and consumers have no idea what is happening to that information.

Not even our laws can keep up with technology and, in fact, if you do business in Europe or want to do business there beginning October 25, 1998, you must adhere to a global privacy code regulating the international transfer of information: no privacy, no trade. No personal information can be sent to countries that do not adhere to this code - including the US. American Airlines is required to delete health and medical information on flying passengers such as dietary needs, disability access and other medical conditions.

Data-mining leaves us with a number of information paradoxes. Your videotape rental history has much better privacy protection than your social security number. There is no federal law forbidding the disclosure of Social Security numbers. But a video-rental privacy law was enacted after a Supreme Court nominee was embarrassed by the publication of a video list that included adult titles. Do I really want anybody tracking my movements through the Melrose Place web page, or the L'eggs pantyhose forum or Victoria's Secret's on-line catalog?

The Fear of Digitized Discrimination Runs Deep

New technologies may facilitate information collection and dissemination, but they also can enable behavior that violates many consumer's sense of privacy. As technology permits us to quietly accumulate knowledge and information about customers, chances increase that we will unwittingly or intentionally violate a customer's expectations of privacy and misuse their information.

Consumer anxiety about the mishandling of personal data is a significant barrier to online usage and electronic commerce. The internet is an environment where there is absolutely no trust. By now most of us know that the internet was designed to survive a nuclear blast that would knock out a large portion of the network, so robustness and survivability, not credit card security, was the most important design criteria. In *The Net,* a 1995 cyber thriller starring Sandra Bullock, she gripes that each one of us has an "electronic shadow, just sitting there, waiting for somebody to screw with it."

Information abuse has serious consequences. For example, a farmer in upstate New York has placed tiny transponders (internet "cookies") around the necks of his 800 dairy cows to enable weighing meters to measure and track how much milk each cow produces. They can identify which cows are making money and which ones are not. The result seriously affects the cow - automatic gates send some back to the pasture, some to the veterinarian, some to the breeder and some to the slaughterhouse.

There is the potential for serious abuses and we have become culturalized to it through movies like *The Net.* There is a deep-seated anxiety by society that database marketing of the one-to-one sort will lead to new forms of customer discrimination as offensive as racial discrimination of the past; a form of customer segregation by personality. How about being branded as a jelly-donut abuser, an obnoxious complainer, or even worse for failing a medical screening? Are we willing to give up this kind of personal information? Do we have choice? Do we think that this will really change?

New technologies may facilitate data collection and dissemination, but they also enable behavior that violates many consumer's sense of privacy. Many consumers are refusing to shop and avoid requests for or are falsifying personal information, thereby corrupting whatever data is collected.

Violate Trust and Destroy Relationship

The fastest way to destroy a relationship is to violate customer trust. Enterprises must respect the privacy and understand the sensitivities of their customers because trust is hard to win and easy to lose. What one person views as an exciting application of technology, another may view as an invasion of privacy. There may also be times when information use violates rules of etiquette without necessarily violating privacy - like Amex's Caller ID folly.

American Express briefly used Automatic Number Identification (ANI) to answer the phones by their customer caller's names, "Hello Mr. Anderson." Call handling times shot up in response to the customer service agents having to explain to callers exactly how they knew their names. The issue was not only creepiness, but a violation of etiquette. This goes to point out that it is at the call center or point of customer contact that this friction will occur.

Companies that offer sales discounts or other perks to "elite" Pareto customers fall victim to a new type of phantom customer who, increasingly, gives fake names and fake information because they feel, in spite of what you want to believe, that taking a few cents off of the price of the product is too low a price to pay for a window on their private life. Ask any data warehouse manager how much information they are hoarding about fake people and fake addresses with fake buying habits. How many of us fill out technical and trade subscriptions with the title of CEO just so we won't get booted off of a free mailing list. In our customer intimacy one-to-one model, many would claim that this is a fair trade; that personal information for personalized service is worth it.

Anyway, It's Not Invasion of Privacy; It's Invasion of Time

This not so much about an invasion of privacy as it is about an invasion of time. George Gilder wrote, "In general the threat to your privacy and the menace of intrusion on your life does not come from true information." (ASAP p. 82, 4/6/98).

An example of an invasion of privacy and invasion of time is the dinner-time tele-marketer who calls to sell me siding because they do not know that I had just built an all-brick home. Invasions of privacy can reflect an inadequate knowledge, not excessive knowledge. Gilder points out that in a sense, telemarketers and other advertisers fail to invade our privacy enough. Many customers are surprised by calls for products they want, i.e. the credit card company that calls to verify the funds being spent. Someone whose certificate of deposit matures at Citicorp's bank, now may get a call from a Smith Barney broker pitching a mutual fund investment. Or selling pharmacy customer's medical information to marketing companies that send coupons related to their disorders.

The call center will be the focal point of this exposure, as they are the locus where undesirable customers are defined and dealt with (a la AT&T automated routing for bills less than three dollars). Interesting abuses and embarrassing stories will emerge: "I am sorry sir, I can't help you today, you were rude the last time you called." Never mind that it was the agent that was having the bad day.

Humanware: The Gold Collar Worker

Since we know from practical experience that greater than 60 percent of the cost of running a customer service entity is in the cost of people and call center staff. There is a discernable shift happening among technology vendors to become more sensitive to the customer service representative (CSR) end of the call center equation. Seeing that the switching technology vendors slowly shift their focus to the application of technology to CSR work quality issues and not entirely on just improving transactional efficiencies has been noticed. The market is very interested in any business message that talks about fairness in agent utilization and reduced turn over in agent positions.

It is too broad of a topic to be discussed in depth here, however a driver of the increasing numbers of applications in the market focused on "agent fairness." Creating "agent burnout" is awareness that powerful Automatic Call Distribution (ACD) technologies foster the potential to turn clusters of gold-collar workers into white-collar sweatshops. Certifications and professional organizations built around the emerging professional breed of super-agents, gold collar workers whose skills and knowledge will require enterprises to pay them possibly more than they pay their managers.

Agents are Where the Next Envelope will be Pushed

The next step in the intimacy evolution does not stop at gaining lifetime-long customer intimacy, the next great gains will come from the agent; the humanware side of the technical intimacy equation. Subtley, but increasingly, the observation is being made that we are burning agents up, exhausting their supply to call centers. It is no secret that turnover, ranging anywhere from 15 to 300 percent per year, is a major problem in call centers. Now agents are smarter, they are needed to be smarter, they are organized and they are professionals. It is with the agents - the new generation of the gold-collar worker, that the next productivity envelope will be pushed.

Best Agent, Best Customer; Flash Point of Productivity

Bringing your best customers together with your best agents can create a "flash-point" of productivity far greater than what traditional customer service technology has offered. In time, we may see productivity improvements of greater than 25:1 or as much as 50:1 when the Pareto customer is matched with the best agent. The point needs to be made here that the "best agent" is not the same thing as the "same agent."

This isn't exactly skills-based routing, which itself has enough longevity to prove that, as a function if implemented incorrectly, actually costs you more in "agent burnout" than it creates customer satisfaction. In today's call center service world, if I have an agent that takes on more skills, I penalize them by burning them up because they are utilized more then someone with fewer skills. With skills-based routing in the long term, the agent is essentially penalized for becoming more valuable.

You Cannot Script Intimacy

It is the agent and the new subjective, decision-making skill-sets they will have that are the most critical piece in the whole one-to-one equation. Think about the Saturn Dealership's service manager and the jelly donut scammer from the context of an agent and caller. Customer intimacy requires alert, sensitive dialogue, but scripting can dumb down an engagement. You can not script intimacy very well, because the agent, as a gold-collar worker, has evolved to the point where they can no longer be considered an just an interface; they're a decision maker now. In fact we approach a point where the data from touch-points becomes quite subjective. Don't tell them how to do it, tell them what you want them to do, then reward them for doing it successfully.

One to One Management-ware

Part of the humanware evolution, is obviously a management-ware evolution. One analyst at Advanstar's Call Center 98 show in Dallas, Texas estimated to me that by the year 2000, 25% of today's call center managers will be gone. They will be replaced because of their lack of adaptability to meet the technology challenges of the next century. There was a clear message to this population of managers: re-skill or perish. Many of these attendees had their antennae up higher than ever, since they realize that they must learn, adapt, re-skill or lose their jobs to - God forbid - a data guy.

Because of the "whack-a-mole" perception of technology "popping" up all over the call center, it takes a different kind of constitution to complete a CTI based intimacy system and applications. The increasing telephony-grade robustness of networked and voice enabled PC platforms is emerging as a real threat to the traditional telephony skill sets managing the ACD platform.

The defining characteristics of most customer service managers are that they operate within the limitations of one-year budget cycles, they are graded on transactional efficiencies and their predominant day-to-day issue is not technology, but human resources. Since call center managers do not, for the most part, operate in time frames of greater than one year, they are looking for incremental and tactical "proof of concept" implementation of applications and technology. The vendors with successful "booth hits" were the ones that could demonstrate proof of a product or application concept in less than a single year's budget cycle.

Being efficiently incubated in the wings of the call center and customer service industry is the Call Center Advisory Council industry certification initiative. Many of us, vendor and end-user alike, recognize the benefits of having a standard industry professional certification, both for managers and agents. Many of us agree that this would go a long way to transforming what we do from being just a job to being a profession. Information and progress reports can be found at www.callcenteru.com. The Call Center Networking Group is another executive resource that could have impactful use in the day to day operation of the customer service center. They can be reached at www.ccng.com.

For the most part, all of this is a moot point because, except for those few call center managers, executives and directors with executive-suite aspirations on their minds, and most call center and customer service focused executives are - at the end of the day, going to return to their call centers and get back to dealing with transactional efficiency issues.

Improving Your "Like-Ability" Quotient

If you want to change your company, change how it relates to its customers. You need to build a communications pipeline that forces to the surface the bad and the good stories from your customer base. Increasingly "like-ability" means insuring that people inside the company are attuned to what people are saying on the outside.

Creating a corporate culture that truly supports customers is perhaps the most difficult thing to do. According to the PW Coopers report mentioned at the beginning of this chapter, perhaps this is a contributing factor to why only 6 percent of companies accomplish sustained yield management status. Management change is tough sometimes, and sometimes it is just not worth changing at all.

Initiatives to improve your enterprise's "like-ability" quotient start in the call center because, even though the call center still suffers the stigma of not being as "sexy" as other parts of the enterprise, the call center represents one of the last great opportunities to differentiate yourself and set new levels of service expectations. If you have a hard time convincing executive management of the value of your call center; consider this: it cost the United Parcel Service, UPS, some $1.2 million for a 30-second television commercial during the 1998 Super Bowl. At the same time, UPS received approximately 70 million customer telephone calls last year, each lasting an average of four minutes. Where do you think UPS's better media buy is? With 70 million four-minute opportunities or a single thirty second Super Bowl commercial?

Be leery of the often fatal results of seeking the Pareto customer. The lure of a few percentage point change in customer retention can result in millions of dollars loss to the enterprise. As with all computer telephony integration implementation projects, leave ample room for heroic failure. Return on investment and information stories like UPS's, combined with the narcotic effect of Newtonization, is one of the reasons for the dismal failure rate of CTI applications. These projects are often rushed into with little or no strategic planning or scheduling demanded of any large technology integration project.

The Petty Tyrant and Champion Sputter

A call center essentially performs a huge, virtual middleware job. This requires the connectivity to new sources of data and information throughout the enterprise. The vital information of customers is most often found accumulated in silos, legacy systems and private tyrannical departments. Mentioned earlier was the fact that the value of information about customer transactions is now more valuable than the technology it resides in has not been passsed up by a new breed of executive - the data baron. Data barons no longer fight for the prized real estate of the corner office, their turf is the information on the server in that corner office. Just going out to fetch some information from a petty tyrants database to populate an agents screen is not as easy as it would appear.

The silver lining to the problem of every call center's petty tyrant is the fact that, because of customer issues like privacy and convenience, much of the petty tyrant's data is corrupt, inaccurate, incomplete or stored on incompatible media or formats anyway.

Lasting and impactful information technology revolutions have always occurred from the top down. Champion sputter is when a project's sponsoring executives start ducking out of "on-track" meetings. Reality looms, interest wanes and the entire project sputters to a slow, grinding halt.

In the absence of leadership, the best advice for call centers is to "just do it" with selected types of technologies, not everything that is in the arsenal. When implementing applications, the call center must focus on the real needs of its callers with practical conciseness and consistency, not simply showcase every possible technology available. Using voice recognition as an example, an enterprise instructs callers to either press "one" or "speak one" ("one" as in a menu selection), but it should specifically offer one way or the other, not both interface options.

Camels and Caravans

Enterprise customer relationships can not change when "the camels at the back of the caravan get the beatings while the one's at the front hold everything up." (Ethiopian Proverb). The core competency of call centers has shifted and call center executive management can no longer value transactional efficiencies over customer intimacies. Those that survive into the next century will be the ones that hone their skill of connecting the dots between the tactics of transactional efficiencies and the strategies of enhancing customer relationships.

If you do not take ownership and start buying into these new solutions, applications and technologies (such as proactive outbound calling, callback messaging, predictive dialing, webmail processing, self-service web and telephone applications, and IP telephony), someone else in the enterprise will. Unfortunately, as your traditional telephony-based vendors rapidly gain data world experience, they will increasingly circumvent you and sell directly to the enterprise CIO/MIS department because they are the ones that "get it."

Newtonization: The Provocative Glamorization of Technology

Newtonization is the glib, topical and provocative glamorization of computer-telephony technology that is far too sophisticated and complex for the users that put it in; when plug and play is in fact plug and pray.

You don't have to know who he is, but Harry Newton is the classic proof in modern day mass media of how you can create an industry on marketing hype and magazines alone (note: Harry Newton recently reported that he sold his CTExpo tradeshow, *Teleconnect, Computer-*

Telephony, Call Center and *Imaging Magazine* empire for an eye-popping $130M and split it among several partners). Miller-Freeman's (who bought out Mr. Newton) huge shrine to the incestuousness of the computer and telephony industry has proven to be just fun and entertaining replete with stylish exhibitor "booth bait" and "show goodies."

The problem with these magazines is that they editorially portrayed CTI as an easy "application kill," but in reality many of the implementations of CTI applications were dismal failures. Statistics about CTI implementations demonstrate a nearly 90 percent failure rate of those that are put in without a vendor, a systems integrator or a consultant. I would consider one of the traits of the very evolved 6 percent of PW Cooper's survey from the beginning of this chapter, was that they probably used a consultant, systems integrator or the equipment vendor. Newtonization is about the over-simplification of customer service technology and communication and the de-evolutionary influence on the entire customer service and CTI food chain. Enterprises now have been "Newton-ized" enough to know from experience that, if they can not keep up with the blistering pace of announcements of new platforms, applications and services, then they had better find somebody that can.

Big boats turn slow, but even the largest of the traditional vendors are showing major evolutionary shifts. A new generation of thought leaders is emerging that is beginning to gain deeper understanding and appreciation of the day-to-day functional realities of the call center. Look for these issues and concerns to increasingly have a greater reflection in the vendor's services, applications and solutions. While it's good to see new product developments and application innovations, the typical customer would like to see less "PowerPoint" and more practical. In either case, neat graphics, screen shots and persuasive presentations belied the fact that few of these companies had actual implementations of their showcased products and even fewer were overtly oriented toward practically addressing the technical requirements of one-to-one marketing. This is the same challenge that exists in the consultant community.

While the consultant community may have plenty of great ideas, if they cannot also specify how to make them work they are not very likely to garner support from the nuts-and-bolts transaction manager in a call center. Like the vendor community to a certain degree, there is a consistent tendency for consultants, particularly if they deal on the strategic level with enterprise executives, to disregard and neglect the practical nature (and implementation requirements) of call center managers.

This technical intimacy stuff is hard to do. With the proliferation of technology and applications, the need for competent guidance - "whack-a-mole" experts - just to pull it all together is becoming more important, if not critical.

It's Holes, Not Drill Bits

It is harder than you would expect to pin down the right number, but the Industrial Machine Tool Distributors Association says that there are tens of millions of drill bits manufactured every year. In the middle of trying to find out the exact number, I read a brief editorial that pointed out that people really don't care how many drill bits that are made, what they want are the holes. Computer Telephony Integration, as an entire technology industry, has been duped - "Newtonized," by the drill bit. Customers want solutions, not drill bits.

One of the opportunities many of the vendors are missing is doing a better job of relating their particular product strategies to messages of one-to-one marketing. And, while the application of one-to-one marketing in call centers has its challenges to the end-user community, the vendors could all do a better job of synchronizing their messages with how this one-to-one stuff is actually being done.

You can give away all of the drill bits you want, but don't forget that what customers really want are holes. Too much technology can turn off customers. Systems and interfaces that offer functions users don't want and lack qualities they find important can generate resistance and frustrate users. IVR is a good example of a technology which often generates a suspicion of evasive antagonism.

Legacy Hesitation

The current vice president of the United States, Al Gore, is a champion of the Information Superhighway's bridge to the 21st Century. Now here is a man that has probably not driven his own car in years. So, from the comfort of his air-conditioned Vice-Presidential Suburban, he has optimistically forgotten to calculate in the condition and effect of bridge fright. Bridge fright is the phenomena that many of those who regularly drive on crowded city-highways recognize as when everybody instinctively slows down as they approach a bridge or freeway overpass. Since they can not see to the other side, they, the herd, instinctively slows. It is the same thing with technology.

While Vice President Gore may be able to see clearly over that bridge into the 21st century, the rest of us are driving at pavement level, and we are not about to cross over so fast. Since we can not see over and beyond that bridge to the 21st century, we remain pragmatic as a culture and we will continue to cling to our legacy technologies and interfaces. We just don't change our user interfaces as quickly as the mass media would have us believe.

Don't underestimate legacy hesitation's influence. User interfaces, operating systems, the telephone, wireless, Windows 95 and UNIX are examples of players with underestimated longevity. Martha Rogers tells people that in ghettos, television penetration is greater than telephone penetration. "If these are your customers, maybe you should learn to talk to them through their televisions," she asks.

Telephony Guys With Laptops or Data Guys With Trucks?

The company's wide area network (WAN) or your cable TV may go out for hours at a time, but when the telephone network goes out, as during the dramatic 26-hour AT&T failure in the spring of 1998, we bolt upright and take notice. The inconvenience caused by no phone service is apparent. No time for downtime is the greatest of the customer expectations and the company that provides this has a tremendous competitive advantage over the ones that don't. From a customers perspective, there is no such thing as service that goes down for regularly scheduled repairs.

On the rubber chicken road-show tour, a frequent question that is discussed is who are you going to buy this customer service technology stuff from? From telephony guys with laptops or data guys with trucks?

Meeting the dial-tone grade expectation of reliability is one of the issues that will keep the "data guys" from taking over the call center anytime soon. As far as reliability is concerned, they have not yet caught on to the call center axiom that "downtime is an emergency." Because of the 100 years of institutionally embedded experience with zero-tolerance to downtime (allowing seconds per year at most), I suspect that it will be from the telephony guys with laptops that the technology of service expert will emerge.

The Peripheral Edge: Columbus Wasn't That Brave of a Guy

History is obviously a little more complex than this, but essentially all Christopher Columbus did was observe the masts disappearing over the horizons. It was obvious that Columbus wasn't a genius, he just could see what was coming over the horizon. This was the simplified version of how he figured out the world was round but, like Al Gore too, the predictive observation of technology is fairly straightforward and it does not take much (well, perhaps a lot of study) to triangulate in on the obvious.

There are several indicator trends and technologies that are more important than others on the horizon. For example, every customer call to Federal Express costs FedEx between $2 and $3 - in fact FedEx loses money on the package. But, when you go to their web site and track the package yourself, they make some money. Another trend example is from a company called Digital Worldwide Services which provides technical support for 14,000 products from 1,300 companies and handles 5 million service request calls a year. DWS resolves three-quarters of their customer orders and queries within one hour.

Look at What the "Net Dragged In" - In a "Net-Shell"

There are plenty of more credible resources available on the internet than here, so I will be limited in my constructive observations and let the reader infer their own conclusions by only pointing out the following statistics.

Business on the internet is growing twice as fast as the rest of the economy and doubles every 100 days and $300 billion in business will be transacted over it by the year 2020.

35 percent of all commerce by 2005 will be over the internet; that still leaves 65 percent to something else, and that is the telephone. Not to mention wireless is growing at 35,000 new subscribers per day. America's public telephone network still carries 40 times more traffic than the global internet.

At any one time, 15 to 35 percent of internet traffic is from adult content sites. In case you are wondering how this could be possible, consider that this book may have somewhere around 10 million bits to it. A video transmission over the internet consumes about 10 million bits in four seconds. So while your company may be sending thousands of emails, it is no match to the bandwidth demands of a few good seconds of video.

Data may be growing faster, but people are still talking. And, in general the end game will be about the touch points, not data points - It's the interface, the touch-point that we should most observant of.

Conclusions

In today's hyper-competitive world, one in which enterprises engage with their customers through a variety of media, keeping customers loyal will demand heightened attention and greater focus from operational management on new strategic business rules rather than old, traditional efficiency measurements. The future of the competitive, successful enterprise is not in the struggle for obtaining and exploiting information, it will be in the cultivation of life-long customer relationships. But let's be pragmatic here and not forget that we, the call center managers and executives, are still captivated by the goal of providing the best possible experience at the least expense to the enterprise.

Reliability, speed and intimacy are increasingly being measured, not by the number of calls answered, but by the quality of the interactions. As a result, many call centers are beginning to see their performance measured against more strategic metrics, such as customer retention, customer satisfaction, consistency in service level, application of best business practices and ultimately, revenue growth. The border between getting a customer, keeping a customer and growing a customer is blurring.

Seeing who represents the greatest profit potential, most likely a customer we can call the "Pareto Complainer," can only be achieved by considering the cost and revenues of customers throughout their life cycles. Customers who, by the way, are getting older and living longer. Calculate that customer behavior will change over time as well.

The cost of winning back customers gets higher every day. Better customer service leads to better customer retention. Successful enterprises will adapt to the new paradigm of business drivers; the customer and their time. What technology is providing, be it through the telephone, internet or skills based routing, is giving people more time to have contact with the people that they want to have contact with, while eliminating non-important, merely frictional encounters. Your enterprise's customers wouldn't tell you this if it was a problem, because they are probably long gone by the time you get the message. The message is clear. Do not waste your customers' time by not communicating meaningfully; you had better have something to say and do it quickly, reliably and knowledgeably. Be intelligent, understanding, and, most important, responsive!

The skills required to manage the Pareto complainer are not trivial skills. Accomplishing technical intimacy is much more challenging than most of us ever expected. The blistering pace of technology innovation isn't just happening "out there" but in call centers as well. If you intend to be a call center manager into the 21st century, then there needs to be less avoidance and more ownership - which is job security. It is easy to be intimidated by non-telephony technology, but the call center is the appropriate place to take ownership of these new media and applications because the call center is the appropriate place to apply the expectations of service levels (downtime is an emergency). The call center will be the focal point - the front lines of frustration where customers are defined and dealt with.

Technology now can provide a foolproof memory of customer interactions, needs and preferences. Once you have been mined, you have been branded for life. Some of the evidence is found like the type of studies by PW Coopers mentioned at the beginning of this chapter. The evolution from farmer to rancher to sustained yield manager coincides nicely with the observation that the most successful enterprises are making the shift from transactional efficiencies to customer intimacies. To adopt these new technologies and the customer expectations they create requires substantially more vision than seeing beyond the time frame of a one-year budget cycle. Take ownership of the Pareto complainer before the wrong department does. In conclusion, management need more strategic vision or they will lose their jobs while the traditional telephony-oriented vendors need less "PowerPoint" and more tactical solutions ,or they will all lose their customers to others that are faster and nimbler.

Chapter Six

Call Center Humanware: White Collar Slave or Gold Collar Worker?

The following was a question asked of Marilyn Vos Savant - who claims to be the world's smartest woman - in Parade magazine, published Sunday, November 6, 1998:

Q: Why do big businesses keep buying phone systems instead of using real people? I am positive they do not save money. All I can guess is that the phone companies have the best sales people in the world. - Jerry Sobotta, Hermiston, Ore.

A: There are lots of good reasons, but I think the best one of them is automated phone systems can free those real people to do something far more productive than repeat computer generated information like your check account balance, flight arrival times or theater show-times. This is how automation helps society progress. We lose a little, but we gain a lot.

Do we? Do we really gain a lot? At what expense?

There is a sublime fear expressed in this exchange that most likely only a few customer service chain executives and managers will recognize. In spite of the fact that an overwhelming majority of the expense of providing customer service is with the humans that provide it, it is the technology we use that dominates our attention.

There is an extraordinary amount of focus in the call center on the "plumbing." Many enterprises do not realize that they (and the "plumbing" that supports them) may be setting themselves up to be branded, chastised, accused or ostracized in the global business environment of dangerously exacerbating the problems of industrialized mass production in the white collar field. We can even find recent written parallels of the "Orwellian" potential of the "perfect prison" described in the Panopticon. (Is it any wonder that, in fact, some of the more successful call centers are staffed by incarcerated inmates of our federal and state prison systems?)

The overwhelming tendency for enlightened customer service chain executives today is to believe that call centers are the battery farms of the information age. They envision and realize that their call centers are a pure source of the digital fuel of the information age - the digitized dialogs that describe and define how customer's needs and preferences are evolving over time.

But, are call centers really battery farms or just factory farms? Are call centers the new sweatshops of the information age? Call centers are emerging to be the most remarkable employment opportunity of the information age. Likewise, there is growing controversy over the quality of jobs at call centers. We can not escape the growing comparison between the high pressure work of call centers and the production pressures of factory farms that make soccer balls, shoes and clothes. The call center phenomenon dramatically demonstrates that changes in technology do not always produce the predicated or intended effect. Better telecommunications was supposed to release thousands to work from home. Instead, we have developed the huge telecommunications factories called call centers.

Technology, in many instances, has redefined the workplace environment from sweatshops to robotics. Computer Telephony Integration (CTI) is the robotics of customer service. However, enterprises cannot script the intimacies required of the one-to-one relationship revolution. You cannot script or flow chart customer intimacy. To the very human customers of an enterprise, the call center, the telephone conversation is the technology. Somewhere in the accelerated spin of Moore's Law (the doubling of computer processor power every 18 months), the demand for service from real, live humans takes a vigorous turn off-track.

Customer service chain executives consistently make the mistake of assuming that the insistence for human contact by customers is a resistance to technology. How much productivity and business has been lost to an enterprise by people who refuse to leave voice mail messages? Our culture has even coined a sufficiently popular truism called "voice mail jail" to support this experience.

Today, many customer service chain executives fail to account for the value of a proven, efficiently working call center. Unfortunately, the next evolutionary leap will have to come in convincing them of the underestimated value of proven, efficient and highly productive humans, the agents that work in them. Agents seem to have become a forgotten lot, not only from a quality of life perspective, but from the position that they are the ambassadors of their organization; they are the company. Making sure that they always have a fair and efficient environment is important and this is not always the case.

The glitz and hype of technology today has created a misdirected focus that is missing the human agent component and how that agent is applied against the goals and strategies of the enterprise itself.

Today, even the venerable management consultants and business seers, Dr. Hammer and James Champy (who co-wrote the book Reengineering the Corporation: A Manifesto for Business Revolution) readily admit that they forget about the people part of the reengineering process that has guided business change for the past decade. Reflecting their influence from an engineering perspective, they agree today that they "were not sufficiently appreciative of the critical, human demand" of the enterprise.

Today, not only does the enterprise need to track and react to the needs and preferences of its customers, but it needs to do the same with its agents. The new labor resources of call centers will quickly emerge as having far different needs and preferences than any generation before it. We are seeing the emergence of the fast-finger-twitch Nitendo generation of worker who has a completely different outlook on work, challenges and expression. The challenges of dealing with Generation X-ers, who have no problems changing jobs as careers and opportunities present themselves, dealing with single parenthood, increasing child care expense, elder care issues, loyalties and leisure are certainly unique, but they are not insurmountable and they can certainly be exploited to achieve the goals of the enterprise.

One of the real dangers of call centers in the information age is that enterprises compromise their relationships with their agents in pursuit of their relationships with customers. Companies can identify customers, but only people can identify with customers. The most powerful computers in the world are no substitute for the power of an intelligent conversation between an agent and customer. What's most powerful about computers are their human parts.

The definition of the new generation of call center agents will be one that can turn an incoming call from a confused, upset or disgruntled customer into a rich stream of impressions, ideas and data. "Deliver a mediocre presentation and you risk the chance of being toned out, deleted or ignored as the consumer searches his or her voice mail, email, or mail for more interesting items." (Angela Karr, Teleprofessional, Oct. 1998, p.8.) In a dialogue-based relationship, the result is a permanent conversation about quality, performance and standards. The successful enterprise will need to have the most talented people at the right place and at the right time. The service agent is the critical component for managing the customer "touch point." Most customer interactions with the enterprise have a human component that remains largely unaffected by technological innovation.

A good example of this is found in saying that a "lowly" call center agent certainly does not know more about a product than the engineer who built it does. But that agent most certainly knows more about the customer than the engineer does. However, too often it is the engineer that drives the product or service issues of the company and not the other way around. This is particularly acute in technology companies with help desks and it may be more important than anything else. The evolution that we are seeing is defined by the shift in awareness that agents and the communicative functions they provide may be more important than developers and marketers of the products they support.

More and more enterprises have reached the same conclusion that customers have become so important that they don't just want people talking to them, they want their very best agents in the world talking to them. Computer companies, for example, should adopt the philosophy that they don't just make high performance computers, they make high performance customers.

For most people, interacting with a person is a more fulfilling experience. A human "hello" is promised by an Allstate Insurance Company office in Sedona, AZ which runs ads in the local paper that assure callers of "an actual, real live hello no matter what time you call us." After normal business hours the calls are transferred to a call center in Nebraska.

Human beings, in the end, execute the work and, as a result, they are at the intersection of the demand for service and the expense of providing it. While we know that the expense of call center agents ranges from 60 percent to 70 percent of the cost of operating a call center, they are more than 90 percent responsible for its success. There is an effective threshold to be balanced between wasting money on overstaffing (the inefficiencies of idleness) and when the availability of too few (or even too many poorly trained) agents exists, customers become dissatisfied, lower their expectations and perhaps never call back. Where is the greater expense to the enterprise?

Many companies say they are obsessed with customers and many companies promise to delight customers - but companies do not help customers, people do. The quality of customer service can not, by default, exceed the quality of the people who provide it. It is with the agents of call centers that the customer service chain strategy must start. The challenge will be maintaining and preserving humanistic functions, actions and the fundamentals of good customer care such as open communication, positive attitude, courtesy and understanding. Today, we face the challenge of infusing and extending all of these humanistic attributes into the electronic realm as well.

In the past, the call center agent was viewed as a low paid, clerical person that primarily just answered the telephone. Today the call center agent is evolving to someone quite different. Increasingly, we are seeing in the enterprise fewer and fewer subordinates - even in low paying job functions. Ideally, call center agents are empowered, incented, motivated, and positioned in learning environments dedicated to meeting customer expectations and the business objectives of the organization.

An increasingly popular description of the new information worker is "knowledge worker." A distinguishing characteristic and the very nature of a knowledge worker today is that he knows more about his job than his boss does - or else what good is he?

Who Should Take That Call?

The concept of the "universal agent" is the Holy Grail of the call center; an agent who can respond to all of the caller's questions and requests as well as proactively recommend new goods and services. Contrast this to the majority of call centers, which are small (under 25 agents) and support small companies. In small companies the customer is always first - not that this is not true for the largest companies, but how often in large companies is the enterprise infrastructure architected to support enabling the lowest clerk to the CEO as a potential call center agent?

This means that at some point in time or for a particular customer, in the right setting, every employee, including the CEO may be called upon to respond and answer the phone (or communication). For example, if there is a big business deal being developed and the potential partner calls the enterprise and the partner is identified by his account number, IVR identification or ANI, and he was calling for the senior VP of sales, the question is, "Do you want that call to go to voice mail or to their assistant?" If neither is available, wouldn't you really want that call to go up to the CEO? And imagine a CEO that responds by saying, "I am so delighted you called, how can I help you?" At that point in time, the CEO has become a call center agent.

We can broaden the definition of "informal agents" when we consider employees outside of the enterprise's mainstream call center as described above. Everybody in the company, even though there may be a formal call center whose main function it is to take calls from customers, is an informal agent. To a customer seeking a conversation, there is little difference (other than perhaps the immediate result) between the shipping clerk at three o'clock in the morning taking a call and the CEO.

Reconciling With the Emerging Demands of Satisfaction

A 1996 Harvard Business Review article pointed out that U.S. companies lose one-half of their customers every five years. Customer dissatisfaction is at an all time low, running at an unacceptable level, with more than one call in ten leaving the customer feeling irritated, anxious, frustrated or furious, particularly in the media, insurance and utilities industries. As e-commerce is driving new levels of richness in shopping online, however, personal service is continuing a slow downward slide globally. These statistics are most likely quite related.

Only 32 percent - less than one-third - of customers with complaints were completely satisfied with agent responsiveness in a recent survey conducted by Hepworth & Company, Ltd. This is an important metric because customers who are partially satisfied are far less likely to make repeat purchases from the company.

So why don't companies recognize the dismal levels of service customers believe they are getting? There are two reasons. The first is institutional denial (not covered here). The second is that only 30 percent of call centers actually measure customer satisfaction (not covered here either).

One emerging strategy for limiting these defections is to reduce customer hassles by making it easier for customers to deal with your company. Customers want to complain; you should make it easy for them. Since call centers represent by far the largest customer touch-point, improving access to agents, information and support services to enable the customer complaint and its resolution.

The challenge is that the bar keeps getting raised in customer service by companies and enterprises that in the past would never have had an influence on your business. Disney is one of the best queue managers in the world. These are not the queues found in their call centers, but the queues for the rides found in their theme parks. In a direct or indirect fashion, the same intrinsic expectation for efficient queuing is applied to the call your customer makes to your call center. People do not compare you to other software, payroll, manufacturing or any type of company. They compare you to where they last got best-class service from such companies as FedEx, Fidelity, USAA, or Disney.

Customer expectations today are such that they want resolution the same day, not sometime. You can no longer tell people that you will look into it and get back to them in three or four days. Customers would laugh at you today for this type of perceived inefficiency. The customer and his time is emerging as the most significant criteria as to whether a customer does business with you or not.

For the past twenty years, delivering customer service by telephone has remained largely unchanged. While voice response technology has improved call centers by reducing live agent calls, its use has been limited to relatively simple transactions. In fact, call volumes are increasing and transactions handled by call center agents are becoming more complex because of the increased information that is available to callers through mediums such as the web and fax back. Longer and more complex transactions continue to require live agent assistance and the demand for live agents has continued to increase - which is different than the number of calls.

The average representative handled 45 fewer calls per week in 1997 than in 1996 (305 vs. 350), but the average call was more than a minute longer (6.26 vs. 5.14), said a poll by the Consumer Support Group. The percentage of calls resolved in the first contact remained at 76 percent and the percent of calls resolved on follow-up dropped one percentage point to 17 percent.

According to The Gartner Group prognostications, shortly after the year 2000, 35 percent of all commerce will be electronic. The ignored irony to this trumpeted expectation is that this still leaves 65 percent something else. For many, and for more time to come, the telephone, either through IVR or with an agent, will remain the more significant and the more preferred approach.

If you look at a typical customer service chain blueprint, there is an area that surrounds the call center customer service applications called the "customer access interface." Basically what the blueprint defines is that regardless of how the customer engages an application, be it an agent, the internet, IVR, telephone, fax or email, the result should be the same - but the result is rarely the same.

The reason that the verbal phone conversation will not go away any time soon is because the customer speaks volumes in every verbal way. The call center environment will remain the most cost effective way to gain insights about the customer's behaviors, emotions, motivations and how his needs and preferences are evolving over time. An IVR or a web page simply cannot accomplish this. When talking to an agent, what determines a customer's satisfaction level or excitement level of that call? The quality of the interaction can be, and often is, very subtle. There are just not many tools that can make a bad agent into a great agent. If you can drive agents up a couple of notches with the proper interaction systems and support systems, it is amazing how much impact you can have on the productivity of the call center and rate of customer satisfaction.

For this reason, moving customer knowledge closer to the customer - such as in the call center, through the use of scripts, cues, and prompts to help agents immediately recognize and then engage more profitable customers. This enables them to communicate more relevant messages and offers while they are on the phone. Effective communication will require accent-free speech patterns, a clear voice and a sense of enthusiasm. Customers are most satisfied by conversing with an agent who is able to carry on an intelligent conversation. Agents that are reticent, easily embarrassed or self conscious do not work very well. A technical support desk agent whose job it is to provide the soothing voice that gently leads confused troubled callers back to digital bliss of bits and bytes.

However, the pragmatic fact is that accomplishing customer intimacy is more difficult than it sounds. A singular problem is that, despite sophisticated database systems, few agents are adequately trained or motivated to make effective use of them, both in the retrieving and inputting. We are simply not capable, for a variety of reasons (that will eventually evolve away), of making other than a superficial use of the information and the raw data from agents' conversations with customers.

In one respect, our increasing dependence on technology for service, while beneficial in certain cases (i.e., bank account balance retrieval through an IVR), is contributing to the fact that agents have become the weakest link in delivering customer service and collecting customer data. They need to be the strongest.

Technology Agents and Influence

There is a Dutch call center - and they consider themselves a call center - that does nothing but answer email for their clients. Like the convoluted interpretation of the term "call center" (similar to "dialing" a telephone), today one can even ask, what is an agent?

Is an agent a live agent or is it a VRU port, or web application? Agents are lots of things today: video agents, software agents, voice response agents and so on. The availability of a variety of media today means that the enterprise has to deal with not only live agent queuing, but queuing with other media as well. There are various philosophies you can apply to queuing multimedia calls in a fashion similar to the queuing of voice calls. Unquestionably, the call center is really the environment that we have grown up with around deterministic rules governing voice calls (average speed to answer, average hold time, etc.). Increasingly, we will have to consider influencing any type of media with (or without) the same set of deterministic rules.

Bring in email, transfers from voice response, video kiosk calls, web transfers and voice calls as well - all now must follow similar deterministic rules. However, the conditioning factor here is economics.

Say a call center is paying $50k a year for an agent and that agent is conducting an internet style transaction. If I can move the customer onto an internet platform, that employee expense can be reduced. Without question, the internet transaction is cheaper than the voice model of transaction. Federal Express has discovered that it costs them $7 - $8 per telephone transaction, but only $2 per internet transaction. Not only do the numbers speak for themselves, but there is also a growing wealth of experience being garnered by consumers from their internet based service experiences. Compare trying to do a stock quote by IVR versus the internet. If an enterprise can move their business onto the internet then they can satisfy their customers with fewer precious resources. This is why many consider the internet as a "third paradigm."

"CTI will have an impact on the most valuable asset in any call center - the agent. Today's technology is changing the way call centers are managed by providing a positive human impact to create a good first impression with customers." (Communication News, 1/95).

In a complete technology-based transaction environment, and assuming that things go right (the item is in stock), ideally a customer service person would rarely need to intervene. Guaranteed information accessibility is what technology provides, it is with the agents where personalization best occurs. The logical preference is to devote that agent's time to up-selling, dialoging and keeping customers than on performing speedy transactions, which is a data entry function at its core.

Technology based self service offers many benefits to both the enterprise and the customer. Technology has provided call centers with two different self-service tools. However, while the internet has several powerful advantages, not everyone has access to the internet. For a majority of the U.S. population, telephone based communications, including the rapidly growing segment of wireless and satellite mobile phones (multiplying to the tune of 35,000 new subscribers per day) is driving voice as the principle communicative medium for some time to come.

Today, unfortunately, helpful technology is running ahead of the ability of most people to use it. Customer service chain executives to this point have in fact acted somewhat irresponsibly in relying on technology to solve deeper problems. For decades companies have been devoting half of their capital to information technology - what do they really have to show for it? The design of call centers has been influenced by the failure of the simplistic assumption that improved information results in increased productivity. Worldwide, companies spend hundreds of millions of dollars annually on technology upgrades to existing systems, new electronic hardware and software, sales and lead tracking applications and various other technologies peripheral to the call center, but do we ask often enough how much is spent on the human resource side of the equation?

Electronic support offers no relief from the mandate of being sensitive to customer's needs. If you remove the option of live time support, you need to state response expectations up front. Here it means that you are proactive as opposed to being reactive and it means providing world class, best-of-breed type of service. It means that your enterprise is becoming totally customer centric.

Front-line service people are taking more calls as a result of technology and a single "point-of-service" contact person may need to access as many as 40 or more separate systems or applications. The biggest problem addressed by the emerging legions of front-office software is that the very legacy nature of incompatible data systems has prevented the enterprise from presenting a unified face to the customers. Software systems in particular can not talk to each other, so many enterprises see no way of commonly identifying a customer across different systems. It is with data-telephony or Computer Telephony Integration (CTI) that enterprises are striving to harmoniously consolidate all of their available disparate technologies. But it isn't happening as quickly as we would want to believe.

Dataquest found that, outside of IVR, less than ten percent of call centers have any CTI functionality installed - blaming the lack of integration on poor systems integration and support expertise by many vendors. Another statistic provided by Datamonitor indicated that, of the 60,000 to 90,000 call centers in the US, only 5 percent (3,000 to 4,500) actually use

CTI applications other than a screen pop.

The unfortunate result is that the call center application and technology market has been growing at such a rapid rate that even crummy companies can survive. In the emerging age of customer centricism, the demand from the technology-based customer is masking the quality of the available technology to enable it. At some point, enterprises will have to reclaim dialogues with their customers, stop listening to market hype and hold vendors' feet to the fire. The fact is that most enterprises generally use technology in low-risk applications first because enterprises, unequivocally, have a significant tendency to avoid experimenting with their customers.

Successful information systems must focus more on customer relationships and interaction than on the information itself. Enterprises are catching on to the fact that it takes more than a slick router and a sensational server. If you think about what CTI does at its base level, it enables everything around a customer interaction up to and after (and sometimes in between, in terms of transferring), but ultimately it only enables a conversation to begin or happen. From that point on the customer cares little about what happens.

To enable the customer relationship side of the business equation, work-flow process applications are increasingly seeping into the call center environment. However, for now, companies are using these new applications for simple transaction tasks such as tracking customer interactions, and will adapt them later for tactical purposes, like measuring the results of different promotional campaigns or incentive structures. One of the keys here is how well your agent, whether it is a live agent or an automated one, can access the infrastructure of the enterprise to be able to intelligently deal with the customer.

After decades of investing in personal computers and software tools designed to maximize individual productivity, we are now turning to maximizing group or enterprise productivity. Front-office applications can be considered "top-line generators," meaning that they actually increase revenues by increasing agents' closure rates, improving customer retention and creating new sales opportunities.

The Measurements of Sweatshops

Though the danger of the accusation will increasingly become more acute, of enterprises course will deny that they run sweatshops; the agents in call centers fear that the measurement and metric function software found in call centers will evolve into total control made easy. Some of these monitoring tactics may seem to promote "big brother-ish" mentality from the perspective of the call center workforce. Graphical user interface, integrated forecasting, scheduling, daily management, long-term staff planning, team, agent analysis and database administration are all applications that envelope the agent's role.

Measuring in call centers will change. When you deal in the lifetime value of a customer, measurements of efficiencies in increments of seconds don't size up for the task. You can run reports and provide quantitative data, but the way something is said - what a customer emphasizes - can be more important.

As an example of the implementation of the "big-brother" technology, as call centers identify the traits and characteristics of a successful agent, enterprises will increasingly turn to high-tech evaluation and profiling prior to employment. In a fashion, a flight simulator for call center agents will emerge as a powerful tool. These tests evaluate the prospect and his ability to handle a healthy level of tension while learning to navigate difficult situations and avoid mistakes. The underlying theory is that if the prospective agent cannot handle the simulated call center environment he more than likely won't handle the real call center environment. This can help predict those agents most likely to be successful.

Designed to screen prospective call center agents against an existing "successful agent" standard, benchmarks have been developed by examining the qualities of proven and successful call center agents. These new applications send to an agent prospect several hours of carefully scripted computer-generated inbound telephone calls and as the simulated caller talks to the agent, the applicant responds. Each call is scored and compared to a statistical database profile of successful agents. The measurements include typing skills, listening ability and problem solving logic.

Another emerging "big-brother" call center tool is the use of workforce management. This is the ability to track agent adherence to an assigned schedule, historically or in real time. We are now precisely alerted when agents sign in late for their shifts, take unscheduled breaks, lengthen their breaks or spend too much time in after-call wrap up. Workforce management application's intention is to alert to deviations from schedule adherence and to resolve conflicts quickly.

But, while workforce management endeavors to keep customer calls from becoming backlogged and service from suffering, it may be measuring the wrong thing. It is easy to set impersonal service standards such as average speed of answer and average talk time. What will emerge to matter more may be the intangible and un-measurable human elements such as attitude, energy, enthusiasm and intelligence. Customers want to deal with people who are bright, perceptive and sympathetic and they care little if an agent is on time, took too long of a break or takes too long for an after-call wrap up.

This is not to say that workforce management software, applications and companies will go away, it's just that they (most appropriately) will evolve into being the most capable of creating the new measurements required in call centers to track the full extent of the customer's related work being performed, not just the number of telephone calls handled. New metrics of the information age will include, for example, the speed of writing, the emotional satisfaction of the customer or the speed of a response to an entire transaction. When new mediums for customer communication emerge such as video, equally unique and new metrics will arise as well, such as degree of tele-genics.

Not Just Techno-Stress for Management, But Agents as Well

While the advantages of enabling every agent with unlimited access to the entire corporate database (an embellished concept, granted) are attractive, its implementation leaves ample room for heroic failure. It is cliché, but technology and the information it enables are a double-edged sword. And, as much as this sword swings to one side as productivity, efficiency and enablement, it swings as clearly to the other of burnout, information glut and the trap of substituting activity for achievement.

Give agents a browser and they will browse. One of the consumptions of workload productivity of agents as they go "offline" to address these requests is costly and quality levels are difficult to monitor. The most difficult and time consuming issues are locating the customer information file, having to "hand-craft" individual replies and determining the optimal medium for service. Today people have to sift through massive amounts of data to find what's useful and meaningful to them. Never mind that much of the information is downright useless, peripheral or inaccurate.

The average American now receives 3,000 advertising message a day, up six-fold since the 1970's. Office workers today spend 60 percent of their time processing documents. A typical call center manager now reads one million words a week. Don't be misled by the notion that techno-stress results from the call center environment provided for call center agents. They are bringing this stress to work with them. The average agents today are not necessarily barraged by technology at work, they are barraged by technology outside of work.

There is a growing concern that the call center environments that we are proposing, enabling or may already have are contributing to already high levels of "techno-stress." Techno-stress is a very real condition that researchers define as the negative effect of technology on people's thoughts, attitudes and behavior of bodies. Techno-stress affects everyone to some extent because we all come into contact with televisions, ATMs, cellular phones, computers, internet, pagers, email and VCRs.

The danger of turning employees into information junkies is that one out of two people actually crave information and they experience a hugc gratification when they find the information they seek. There are voice messages and fax messages but the internet has a true addiction to it, particularly online news feeds feeding America's fast-food mentality toward information. It is instantaneous gratification and it entices you to use it more often.

Information addiction clogs data pipes, slows down legitimate business access and works against call center and enterprise productivity. Data-holics. Information overload can reduce your attention span, disrupt your family life and impair judgement. Info-glut leads to boredom, stress, loss of sleep and loss of productivity. As call centers become a central indicator of the differences between the environments of gold-collar workers and white-collar slaves, one distinguishing characteristic will be increasing the availability advice for dataholics of cleansing data fasts and data naps - regular breaks away from exposure to information.

Who is this New Agent?

The Nintendo generation is entering the workforce, and they are coming to work in your enterprise's call center. This next generation of worker will transform the nature of the enterprise and the way that wealth is created as they create the new culture of work. The new generation has a level of technical savvy-ness and a thorough set of expectations and values that are quite a bit different than the traditional corporate and call center culture.

The call center and its derivatives, increasingly ill defined by the nomenclature today, represents one of several enterprise functions where we see the significant emergence of the gold-collar worker. What exactly is the "gold-collar" worker? (The term gold-collar worker was first advanced in 1985 in a book titled the same by Robert E. Kelly.) The gold-collar worker is fundamentally a new class of call center agent whose skills and knowledge will require enterprises, eventually, to pay them more than they pay their managers.

What will drive the emergence of the gold-collar worker is the fundamental tenet of economics; the availability or scarcity of resources. Increasingly we will see the intrinsic demands of agents becoming a strong bargaining point because of the demand for these skill sets. Combine the demand for skills (this is why skills-based routing was developed) with the ease of changing employers and the cost of replacing agents and you have a formula tilted to the agent's favor. Unhappy employees can switch companies more easily when the economy is doing well and replacing an agent costs one to one-and-a-half times the departed agent's annual salary.

An increasing number of skills will be required by agents, many related to the internet medium, and finding experienced, trained agents will become more difficult. The more complex the transactions, increasingly found in today's world where customers are increasingly knowledge-ified, the skill sets of agents are not so easily homogenized. In fact, the enterprise must be careful that they do not hire computer geeks with the right skills but the wrong disposition.

They have an ease with technology; digital tools spell trouble with the traditional enterprise and the traditional manager. This generation will create huge pressures for radical changes in existing companies.

The pool from which we will increasingly be pulling our agent resources from is from the so-called "Generation X-ers.

Generation X is considered to be those born between 1965 and 1976. They are the first generation to have grown up with the technologies of information and they have learned to communicate far differently from any preceding generation. Generation X is the first generation, as author Don Tapscott says, "bathed in bits" and that "for the first time ever children are the authority on an issue of importance to society - technology."

Effective management of the gold-collar agents will rely on subtleties, adeptness and deftness to a degree never employed before. Understanding these agents' important issues is the key to avoiding the stinging brand of "white-collar sweatshop."

The decade of downsizing and right-sizing has definitely seen a drop in employee loyalty from where the company came first. Now it is the family; my own personal desires are all taking much more influential positions that say that the employee no longer thinks of himself as a company man as in the past. It is the changing fundamental economics of business that are changing the employee's desire for a higher quality of lifestyle, because this is a higher priority than it has been in the past.

Talented Generation X-ers want no part of companies that deliver conformity, rank and overly controlling practices requiring dress, travel, hours and office recreation, but promise an exciting career with plenty of educational opportunities.

They have a very strong sense of social responsibility and they care about the collective, but they have a strong sense of entrepreneurial risk. Many recent generations of workers have a very deep social sense; their causes include support of education, environmental issues and a belief that technology can solve many human problems. They will demand time for private projects and renewal.

This generation is exceptionally curious, self-reliant, contrarian, smart, focused, able to adapt, high in self esteem, and posseses a global orientation. They are more technically literate but with shorter attention spans. Generation X-ers prefer information that is concrete, concise and to the point.

There is a clear connection between the attitude of it reps, the quality of their performance and the amount of information they gather. Forty to fifty percent of the success of the call center is in the attitude of the agents. The skills needed are not just creative brilliance or intelligence, but also require a certain attitude. An increasing characteristic of successful call center managers and customer service chain executives will be a trait (again framed by Watts Wacker) of intelligent disobedience. Intelligent disobedience is what seeing eye dogs are taught, essentially that they are to obey unless they have a better idea. The act of intelligent disobedience allowing the agents greater decisions making discretion's. Disruptive people are an asset.

There is great value in agents that rely on their seat-of-the-pants impressions of customers (just like the old corner grocer used to) with no hard data to back up their intuition. As the actual dialogues and the preference and needs nuggets they reveal become increasingly important, accurate subjective observations with customers become a critical skill (How else can an enterprise identify and measure a mean customer, for example?).

Today, call center agents are employed based on their verbal talents for customer service and subject matter experience. Increasingly we can expect agents to have internet skills, writing skills, navigating web-sites and experience with computer supported environments.

One of the surprising results of the information age is the momentum that has turned back to the written word. Demand for agent's skills and competencies will continue to evolve, especially writing skills. Writing will become increasingly important, specifically with customer communication. Finding agents that perform well on the telephone and can perform well electronically means that basic skills cannot be taken for granted. In electronic environments, writing skills are much more necessary than talking and articulation. The fact is that good talkers may not be able to develop the succinct summaries and concise statements required when building net-based knowledge bases and customer relationships.

There is real advantage in being able to craft precise written communication. Technology may enable an agent (or his manager) to write and distribute an email very quickly, but it doesn't help much in what to write or how to say it. The problem with the too technical person is with how they communicate. A agent's skill may well be in overusing technical or clinical terms such as, "dollars-per-kilobit-per-second" or "average speed per answer" as opposed to revenue, competitive advantage and lifetime value terms more familiar to his peers. In the technical support environments in particular, agents may be one-hundred percent buzz-word compliant, but need a greater characteristic of not needing to demonstrate it.

I can attest to the conversation with a retail catalog agent who, at the conclusion of my call, asked me if the "n" in Anderson was "n" as in knife! A call center manager colleague of mine also related a story, still used for training purposes, about an agent who asked a customer if, "that's 'q' as in cucumber". Basic writing skills are critical.

Finding the Right Agent in the First Place

One of the meta-trends of the information age that will have a significant impact on call centers is the decentralization of location. Agents, in a few short years, will be engaging customers from the mountains, beaches or even in space. Companies now are beginning to understand that they have to broaden their reach to these virtual resources because the people are out in the countryside now. Technology isn't just making for virtual customers' it's creating virtual employees as well.

In the enterprise we should define virtual as being two things, one being the non-living agent application (not discussed here) and the other virtual implies that there is an organic being located somewhere other than a centralized call center (or highly networked for that matter). One form of virtual is an agent working from home. The next generation won't be bounded by brick and mortar, it will be virtual. The agents can be working from home, they can be telecommuting and they can be working from anywhere.

Also, agent turnover is at an eight year high and the cost of replacing a good employee ranges from one-half to several times a year's pay, depending on the job. Some suggest that high turnover in the industry is skewed by the high concentration of call centers (Omaha, Phoenix, San Antonio) because they poach each other's staff. But handing out raises is costly and works less and less with today's increasingly diverse, life-style conscious work force.

High turnover, wage competition and guarding against natural disasters and network failures are just a few of many influences causing companies to scatter call centers all over the country. But significant also are employee desires for remote connectivity. The Nintendo generation is beginning to realize that they don't have to go to the company, the company will come to them.

The fact is that, although the labor pool and the cost of living may be terrific in Sioux Falls Idaho, there is just not the type of, for example, Java windows NT resources an enterprise may need. The fact is that these very same enterprises can't afford these same resources when they live in a Silicon Valley or the Massachusetts Route 128 high tech belt. In the technical support area particularly there are areas that have the resources available like may have to draw on to serve your customers you have to find those people and contract them where ever they are. Agents do not want to move and they do not want to change an established quality of life where they are, so the enterprise has got to figure out how to secure that resource without having that resource on site, in a chair or desk at its location.

In the old industrial age model, if you wanted a job you had to move to Pittsburgh because that was where the steel plant was. Today, in the information age, we have built interstates and superhighways and the tables have turned. Now enterprises must go to you to have you as the resource. The company needs these employees now, more than ever. This explains why regions in Texas, Oregon and Wisconsin are fast becoming growth Meccas. Employees are finding that they do not want to compromise the quality of life available there and this is what these resources are looking for. They know that they can become telecommuters and they can still find work and be productive employees, making a good living by living somewhere that is very comfortable for them.

Striving to boost productivity, companies have spent billions of dollars on information technology, but as the asset base of many companies shifts from factories to talent, attracting, motivating and retaining quality employees at every level becomes not just a vague slogan but now a matter of survival. But for now, many of these new high tech tools are often left idle by a very old fashioned labor shortage.

In fact, there is a growing, not declining, dangerous shortage of talented people who can fulfill the enterprise's mission. For the immediate future, agents who know your business and can handle customers appropriately will be harder and harder to find. Labor resources will increasingly become scarcer. This critical trend intersecting the critical shortage of information technology workers is entering a new phase. This labor shortage is no longer just an inconvenience for businesses anymore. The shortage is having an obvious bottom line effect on companies particularly in technology businesses.

One of the emerging challenges for call centers will be in finding labor markets that can support the call center. The number one criteria for site selection today is labor demographics (followed by space availability, telecommunications, infrastructure and tax climate). Other factors such as weather conditions and economic development incentives also weigh in on the formula.

Enterprises are finding that they are having to move their operations to more resource available areas, particularly companies that have traditionally been based in the downtown areas of major cities. Increasingly, these companies are concluding that they are not in the train business, or the people moving business. They are moving to where taxes are cheap, there are incentives to develop in other areas and these central locations do not have to support a large labor force.

Ultimately the call center labor supply will catch up with demand. The nurses' shortage of the 1980's, for example, has turned into a glut.

From a management perspective, one of the unspoken threats to the efficiency of call centers will arise, slowly and unsuspecting at first, of organized agent unions. Losing political weight in the realm of industry, organization activity of the information workers will increase under the aegis of guarding against the exploitation of "workers" in the white-collar sweatshops.

To preempt this trend, one of the best propositions emerging from the call center industry in the past years was the idea of an industry certification initiative. Many of us, vendor and end user alike, recognize the benefits of having a standard industry professional certification, both for managers and agents. Many of us agree that this would go a long way to transforming what we do from being just a job, to being a profession. In this initiative would be the type of evolutionary protections against the fears of call centers becoming, or being identified as the poster environments of information factories.

To Train

Many methods have been devised to minimize turnover in call centers, ensure that the workforce is happy, motivated and has plenty of incentives. One of the prevailing characteristics of the Generation X-ers is their desire to learn (although you may not be able to tell it from their school grades). Continual education will be one of the most significant incentives for attracting and retaining the knowledge workers of the information age. The new agents of today need to be trained because that is what they want. Improve workforce skills by investing in training agents on how to make the best use of the technologies at their disposal.

One effective training trend of enterprise is to "cross-functionalize" their employees from one department to another. In the telecommunications business, for example, one of the reasons that service from the RBOCs is so poor is because their agents often don't understand their own business. In these environments (where customer loyalty is different than no choice), all agents know is how to follow scripts.

The definition of "knowledge worker" must expand to include agents that clearly understand their own business and that they can answer more questions about their business and give the customers a better feel for the competencies, services and products in the enterprises. To accomplish this, for example, it will become entirely appropriate to have those working in accounting participate regularly in customer calls. For that matter, perhaps we should encourage the CEO and other executive officers to participate as well. Let them take a call from a customer and see how difficult it is to collect a payable or resolve a technical issue.

For "cross-functionalization" to be effective, call center management must constantly find ways to expose agents to the rest of the organization – and the rest of the enterprise to the agents. There are effective benefits to be gained from interaction through boards, groups, committees and teams where the call center is represented. Referring to the educational desires of Generation X above, the desire for education extends to the benefit of working with other smart peers, teams and leaders. To this end, one of the emerging roles of call center managers will be the ability to create highly collaborative environments.

Beware, like substituting activity for achievement and paving the cow path, teamwork isn't always good. The danger in cross-enterprise fertilization is that it often spares the feelings of those who make peace with shoddiness and imperfection. The last place that an enterprise wants institutionalized poor quality is with the enterprise's relationships with customers. If an agent has learned that it is acceptable to stonewall an enterprise's vendor on an account receivable, then the influence may extend to not aggressively resolve a customer's problem.

Related to this is the overwhelming simplicity of believing, in error, that an enterprise can throw hardware and software (workforce management and schedule adherence being two) at call center problems. The management of call centers will require a more holistic approach. If human resources, product development and IS are not involved in the day to day operations of the call center, somebody ought to lose their job.

Turnover in call centers is high. It's tough being nice day in and day out for weeks, months and quarters at a time. (Teach that to the CEO.) What if they are treated poorly by supervisors or looked down on for doing "phone work"? The most significant hit to morale is when the agent realizes that there is no true career path at the company.

In pursuit of good agents, most enterprises paint positive, glowing pictures of the jobs they are seeking to fill. Many agents start off with high expectations about their opportunities with the enterprise such as expectations of, "you will double your salary in six months, you'll move up to supervisor and you will rarely have to work weekends." Ohio State University Psychologist John Wanous has discovered that enterprises can actually reduce turnover by lowering the expectations of new employees.

He found that using realism can reduce turnover by as much as ten percent. In his studies he discovered that after one year of employment, fully one-third (33 percent) of employees who had been optimistic about their opportunities had moved on to other jobs. This is compared to the one-sixth (16 percent) of those who got cold-water treatment before or right after their employment, that had left their jobs.

Knowledge Managers or Slave Tenders

Although the provacativness of characterizing call center agents as "white-collar slaves" is an intentional embellishment, this unsettling possibility can only be enabled by the future managers of call centers. It is the characterizations of these managers that attention will now be turned to.

If, and when, the call center gains the recognition as the only source of raw, pure and true information regarding how customers' needs and preferences are evolving over time, the role of the customer service chain executive will change significantly. Forward thinking organizations will reconsider and rethink the call center manager role in the context of being "knowledge officers," which has a far broader scope than the view of the traditional ACD, human resource and network manager prevalent today.

Broadening your skill sets and technology application experience will be the only way to insure job stability. The model of the next generation call center executive includes fundamental proficiency with new media all day long to communicate internally and externally to the company. If you and your agents need to know how to write, consider the demands executives will face.

The new customer service chain executives and managers will build their enterprises based heavily on internetworking; striving to create a real time company from real time customer information appropriate for an economy that's becoming increasingly real time. The newer media enables the creation of a flatter, more open company and an enterprise culture that is rapidly responsive. The successful call center managers will be considered experts at creating great value from information flow and we will see their compensation equal the most senior managers, directors and executives.

The executive that replaces today's call center manager will have strong business process knowledge and an ability to think beyond the technology to understand the business issues. Their responsibilities will be defined to include the implementation of strategic plans to leverage information technology to support business goals and not defined by metrics such as average speed of answer or average wrap-up time.

Breaking with the generally unacknowledged job safety found in being the only person who knows how to program the ACD or manipulate workforce management applications, the enterprise will recognize that it no longer has use for or the need for information overlords; rather they are looking for managers who have the combined interest and expertise in technology with powerful leadership skills, real business empathy and an intense ability to listen. An increasingly important skill in the call center manager's portfolio will be something akin to industrial psychology.

Smart call center managers know, for example, that the best ideas, concepts and constructs originate from the floor - from the agents themselves. In fact, astute managers realize that suggestions from agents should be the most valued because they know that the agent is the one with the closest contact with the customer and has the best, if not the only, opportunity to improve customer relationships.

Call centers will no longer be managed, they will be led. The highly paid call center manager will be the one who has developed the personal ability to build call centers, sell them and support them while sorting through a sizable number of problems, every one of which is influenced by the challenges of corporate culture (read: politics). This new breed of manager must be capable of successfully executing against a corporate strategy with a strong end user service orientation and have the proven capability to extend the customer service infrastructure to support every other part of the enterprise - both internal and external.

On the technical side of the equation, a side that most managers still must become proficient at, the new role will include being a system integrator. Instead of assembling components as an expertise, call center executives will become portfolio assemblers, pre-selecting and assembling multi-vendor solutions and then supporting those solutions. The trend that enables the call center managers is the evolution from managers of the bits and bytes of data to custodians of the most significant wellspring of enterprise knowledge.

To adapt to the needs and demands of the emerging generation of agents, customer relationship executives' new challenge will be to put into place the networks that manage the output of these agents; they, in fact, will be managing the processes and not the employees. It will be management by influence and not by direction. These executives no longer will "manage" people, they will lead them. The way one maximizes their performance will be by capitalizing on their strengths, experience and their knowledge, rather than trying to force them into molds.

According to Peter Drucker, what will motivate call center agents and the knowledge workers of the future will be the same thing that motivates volunteers. "Volunteers must get more satisfaction from their work than paid employees precisely because they do not get a paycheck. They work for challenge. They need to know the organization's mission and to believe in it. They need continuous training. They need to see results." If managers have done a good job of communicating the enterprise's mission and firmly believe in every agent's ability to participate and contribute to accomplishing the mission, they are well under way to being as evolved as the agents they manage.

One of the most significant (and appropriate) trends in call centers is to make agents responsible for their own training and development. This can manifest itself in many ways, but at a high level the best call centers identify and communicate growth opportunities to their agents and personalize growth plans. Then they encourage agents to be proactive in achieving their own growth plans. Imagine a call center where agents crave being monitored for their own personal achievement, not for conformity's sake.

Conclusion

More enterprises than not spend a lot of money on recruiting, training and nourishing agents only to place them in environments where they burn out quickly. This dramatically contradicts a new model focused on the agent as a more important piece of the call center than has been traditionally emphasized in the past. The technical intimacy approach to customer-centrisim means that there has to be a fundamental call center change, in fact an enterprise-wide change, in institutionally embedded behavior. You are either contributing or complaining.

In defining what the call center of the future will look like, Kelly Conway, the Chief Executive of the call center consulting firm, Technology Solutions Company, envisioned that, "Agents will find themselves called upon to handle a wider mix of transactions, complaints, service call management, marketing and even public relations tasks. Moreover a work day, which was largely routine will take on a much larger degree of 'emotional labor' typical of this kind of intensive customer contact, the work day will typically become either more exhausting, more exhilarating or both. Most call center will find themselves formulating changes in business rules and processes that allow them more discretion in customer interaction and also will involve more gray areas and greater risks."

Good service is just solving a problem – delivering what people expect to receive. Great service is getting below the surface of the problem and delivering what no one expects to receive. Every problem is different, every call is different. The enterprise can not prepare scripts to cover every situation, let alone every personality on the other end of the line.

The preferred skills of the new generation of agent will be with listening, learning, assessing and refining. Ultimately, great service requires sound judgements from the agents on the front lines. The only way to get better at answering customer calls is to get better at making judgement calls. Good judgement starts with deep knowledge, training. Great service is an art; a good call is subjective - literally a judgement call.

Successful enterprises today must recognize that customers are assets to be managed with a superior service bias. This bias will lead to greater retention and those retained customers will generate more revenues, at lower costs, than new customers will.

Knowledge may be power, but communication skills are the raw transport of the information age. Not being able to speak the language of the business is critical, as will be being able to understand the business issues and the role of the call center and the efficient use of technology tools as it becomes available.

It is important to understand that, for the gold-collar concept to be accepted, customer service chain executives must first convince themselves, their executives and ultimately their customers that it is with the agent that the most significant gains in call center efficiencies can be obtained. Imagine a customer base that has no concept or need to demand to speak to a supervisor because of an unsatisfactory contact.

While the trend has started to change recently, at least at a superficial executive discussion level, most customer service technologists tend to neglect the agent as a crucial and strategic piece of the customer technology picture. It is just easier to know how to program switches than to manage people.

It is in the technology of "humanware," the agents and customer service representatives, that customer loyalty battles will be won. The gold-collar worker is a new breed of articulate, educated, and responsible "agent." The role of management will be in enabling the remote environments they will work from, and the technologies that will empower them to be more effective and efficient. Their lifestyle demands are not the same as the demands of the executives that manage them and great skill will be required to reconcile these differing expectations. Balance will have to be struck between "over-enabling" critical service agents who are simply not prepared to use or take advantage of the expensive information technology resources at their disposal.

In conclusion it will, in all hope, be important to share a perspective contrary to the prevailing hype about the technical evolution of call centers. The contrast between the white-collar sweatshop and the gold-collar worker represents a new concept in customer, employee and executive interaction - a philosophical change associated with the underutilized and underestimated intellectual resources available in the call center.

The intellectual application of the information available at the moment the enterprise is engaged with their customer, regardless of the medium, requires a new way of thinking. Information is the spoils of discovery. Where else can this discovery take place than at the moment of customer engagement and who else can enable this than the call center agent?

There is an African proverb that says, "If you know your history, the future will not trouble you." Changes from the discoveries of science, geography and customer service have required new tools, new technologies and new ways of thinking. Those customer service chain executives who anticipate and understand the fundamental nature of the changes ahead and actively reshape their business models to meet these inevitable changes will be best placed to exploit these opportunities. Those that do not will face a difficult transition from the legacies of the past to the intimate realities of the future.

The emerging technologies of customer service can support the sophisticated processes that in turn support enterprise strategies. But, fundamentally, customer service chain executives and leaders will have to make the key decisions that will shape the nature of their operations – information sweatshop or liberating opportunity.

The big returns are yet to come for those enterprises that learn to use agents - associates - as strategic tools. The enterprise may have the most current of the service technologies and its executives may be the most proficient in its implementation, but, as Nadji Tehrani succinctly says, "Treat good people well and they'll make your technology hum and sizzle like it was meant to. You'll be rewarded by outstanding revenue and enjoy a leadership role in your industry."

The fundamental determination of whether your enterprise is running a white-collar sweatshop or liberating gold-collar workers occurs at every moment that any agent (or manager) asks himself the question, "Am I a human being or a human doing?"

Don't negate the importance of technology, but also don't negate the importance of people. If there is one message about the strategic role of the call center in the enterprise's future, it is to look beyond the current technology based trends and select a long term strategy that views people, human resources, as the primary source of enterprise value. Only then can the enterprise evolve, adapt and change to embrace the new generation of knowledg-ified workers, their ideas, and their different view of work and working.

Chapter Seven
Y2K: The Big Bug
Apocalypse or Annoyance?

Abstract

Never before in history has software's vital importance to business been so visible and vulnerable. The inability of almost every computing device today to properly adjust for the millennial rollover date of December 31, 1999 to the year 2000 will have a far reaching ripple effect from inaccurate, bad data. Older and aging routers will get stuck in an endless loop trying to process the new date and then resetting themselves and forgetting that Y2K is also a leap year. The year 2000, or also referred to as Y2K, will start a ripple effect of bad data - making the business problems caused by Y2K not just a possibility, but an inevitability. Certain industry segments such as banking, finance and health care are more at risk than others. However, in spite of the certain recessionary trends the Y2K problem ignites, certain industry segments will flourish. Customer service applications, and the computer telephony technology surrounding call centers is one such segment. The Y2K bug is forcing enterprises to absolutely rethink their information technology vision. For customer service and the call center technology and applications, Y2K will likely prove to be more of a nuisance than a neutron bomb, creating minor inconveniences for customers and employees downstream.

Introduction

At first, there is no power and dial tone disappears, cars stop abruptly in the middle of highways, airplanes fall out of the sky, medical respirators, heart machines and life support systems fail in hospitals. Then banks, schools and stores close. Sprinkler systems report and alarm on false data or fail to go off. Later, layoffs become rampant, unemployment rises dramatically and then the whole economy grinds to a post-apocalyptic halt and drowns in a dismal digital depression of deprevation. Only those that have taken to the hills miles from the nearest big city stocked with survival gear, six months of provisions, their life savings in gold, printouts from Social Security and TRW, packaged food and bottled water will survive. The crash of 1929 pales in comparison to this crash.

Annoyance or apocalypse? Nothing has gotten more press than the year 2000 problem, also called Y2K – which is in the same league as Godzilla. It's Y2K, The Big Bug, and at midnight, December 31, 1999, as the ball drops into January 1, 2000, many alarmists are predicting the collapse of everything that depends on computer chips and software, from VCRs to nuclear power plants. On January 1, 2000 (or more likely a few days later, because January 1, 2000 falls on a Saturday), people are going to turn on their computers and find that their systems either refuse to perform tasks involving dates or have hopelessly mangled select applications and programs.

The pervasive fear that Y2K will quickly snowball into the most complex tangle ever untangled comes from the strange uneasiness and paranoia growing from being constantly bombarded with Y2K bug articles, books and stories for the past several years. ABC Nightly News reported that experts expect a 40 percent chance of a global recession. Y2K has been on the cover of Business Week and Parade Magazine, discussed on National Public Radio, written about in mainstream newspapers and even lampooned by Dilbert. "Millennium Bug" was voted word of the year for 1997 and Forrester Research claims that the Year 2000 problem has become the most-hyped issue in the history of technology.

To put some perspective on the enormity of the problem, the federal government's Office of Management and Budget estimates that the price tag to avert widespread government computer crashes from the Year 2000 Bug has grown from $2.3 billion in February, 1997 to $2.8 billion in May, 1997 to $3.8 billion in August, 1997 and $3.9 billion in January, 1998. The Internal Revenue Service is expected to spend over $900 million to scour 62 million lines of computer code, second only to the Defense Department as the costliest Year 2000 problem fix in government. This brings up the point that – and it is no different in the private sector – if the fundamental systems of an enterprise are in disarray, the downstream systems in the call center are certainly affected.

Y2K is everywhere. In newer mainframes, in microprocessors-based task-dedicated devices, in PCs and in many other turnkey computer based products – everything in the call center or enterprise telecommunication systems. To get a root perspective of your enterprise's Y2K vulnerabilty, consider taking an inventory of every electrical devices there is.

Y2K problem will be the argest most expensive project ever faced in the information technology age. The decision to save a few bytes back in the fifties, sixties and seventies has led to a technology fiasco in the nineteen nineties. Some companies merely view the Y2K problem as hyperbole and just plain hype.

What many predict will be a digital apocalypse can be limited to a mere annoyance with diligence, planning and preparation. This paper discusses the extent of the Y2K Bug, what enterprises, industries and systems it will affect and what impact it will have on the most visible operation in the enterprise - the call center.

Historically How This Could Have Happened

It's mm/dd/yyyy, not mm/dd/yy. The Millennium Bug is the consequence of computer programmers who, during the first twenty years of computing history (roughly 1960 through 1980), represented dates with the year expressed as two digits like 95 (mm/dd/yy) as opposed to four, such as 1995 (mm/dd/yyyy). This was done for several reasons. At first, they did it because it saved disk-drive storage space. One estimate is that by using the two-year format instead of the four-year format, the typical enterprise saved over $1,000,000 per gigabyte of total data storage in the 30-year period from 1963 to 1992. It is a symptom of programmers trying to squeeze the most performance onto the smallest computing footprint. Saving two bytes per date filed made sense back then, and nobody ever expected these systems would last three decades. And the most significant reason was that they just didn't think to do it any other way.

30 years ago, computers had about one percent of today's processing power. The computation time needed to convert dates to and from integer format would have greatly extended expensive batch processing time. A four(byte integer filed could have stored 179 year's worth of dates. Still, others question the true savings numbers by not using a four-position year 25 years ago. The issue was that the two-digit year was a feature, not a bug, so it made good sense at the time. To this day, many embedded clock/calendar programs output "wrist-watch time" – two digit years. The way we encoded dates was an intelligent design intention during the 1960s and 1970s, when most of the "back-office" applications used today were actually written.

Nothing in the short history of computer and information technology has prepared us for the fact that the Y2K Bug is actually three "bugs." The triple witching bugs of the new millennium are:

September 9, 1999. A sequence of 9s was used to test portions of service applications and trigger events to automatically execute on the date September 9, 1999 or 09/09/99. Many of the newer "software gremlins" popping up are attributable to 09/09/99.

January 1, 2000. The Big One. On this date the two-digit clock codes of uncorrected equipment will turn to 00, fooling the applications and equipment into believing that it is the year 1900 or 1984, wreaking havoc on systems, equipment, interfaces, devices and ultimately people.

February 29, 2000. If the world somehow survives collapse on January 1, 2000 then it will be surprised to find a day missing at the end of February, since the year 1900 was not a leap year and the year 2000 is.

Mainframes have gotten plenty of attention in the media because of the need for programmers who actually know (remember) how to program in the old mainframe languages such as COBOL. It's true that millions of lines of mainframe code must be adjusted if legacy systems are going to round the corner smoothly into the next century. However, no one in the telecommunications industry expected that their systems would make it to the turn of the century.

However, for the 55 percent of non-mainframe systems, there are far fewer tools to fix the Year 2000 Bug and as the clock turns to midnight December 31, 1999, browsers and computer systems that have the Y2K problem will show the year as either 00 or 84. If this date is forwarded into a form or application on a web site, it may cause the root server to engage in an error and stall. Y2K is likely to be interpreted as 1900, instead of 2000, or even such odd variants as the year "-1."

As though the Y2K Bug weren't creating enough of a headache, there is another sister problem related to date and time-based computing processes. The Year 2000 is also a leap year. Even if your systems and platforms can function satisfactorily believing that they are operating in the year 1900, the year 1900 was not a leap year and these systems will account for 28 days in February, not for a leap year's 29 days. The leap year rule in the year 2000: If a year is divisible by four, it is a leap year. If it is also divisible by 100, then it isn't. But if it is also divisible by 400, then it is a leap year. You may be surprised to find that your computers, which may or may not survive the Y2K bug, on February 29, 2000 your systems will start processing that date as if it were March 1, 2000.

Apocalyptic Prophecy; The Ripple Effect

A typical machine, from a car (which has hundreds) to a coffee maker, for example, contains at least one embedded processor that maintains the current date and time. There are time or date-sensitive embedded chips in elevators, thermostats, security systems, fire alarms, irrigation programs, lighting and lamps, door locks, microwave ovens, digital watches, answering machines, PDAs, voice-mail systems, smart telephones and digital cameras, to name just a few devices. Every PC or application server and telephone or station set has a real-time clock susceptible to the Big Bug. It is these embedded micro-chips and programming code, dedicated to a single time-keeping task, that carry the ticking Y2K time bomb, literally.

Increasingly headlines yell "Prophet sees mega-crisis" and "Y2K: Wartime Mentality." The articles say that people need to start stockpiling food and supplies, and withdrawing cash. Traffic lights and subways will stopand the civil unrest follows. Nine months later there is a spike in birthrates called the Y2K baby. Even the elderly are getting headlines in their monthly senior-citizen newsletters – headlines asking, "Will you be ready for Year 2000?"

And, while the satellites that run the Global Positioning Systems (GPS) will fly just fine as they did during the great pager outage of 1998, some transmitter dates will be skewed, causing ships to run aground and planes to land hard. Here is list of benign things that could go wrong because of Y2K: microwave ovens, city lighting systems, air traffic control, telephone systems, delivery of checks, automatic teller machines, elevators, sprinkler systems and world wide recession. Being in an airplane or having your x-ray taken is a very serious issue with very serious and deadly consequences.

Timing is important because it controls the root functions of an information society such as time and date stamping, database backup, best route switching, synchronization processes and other applications. An example of an even more real world consequence of Y2K is that any system that deletes the oldest backup file of a reporting or call-accounting database each time it creates a new backup file will actually be going to eliminate the most recent file. It will think it is actually the oldest – after all, it appears to have been created in 1900 or 1980. A phone call started in 1999 and ending in 2000 might be billed as just over 52 million minutes (60x24x365x99); a 100-year phone call (if the caller was on a friends and family plan at ten cents a minute) is a bill of $5.2 million dollars!

Never before in history has software's vital importance to business been so visible and vulnerable. The Y2K threat has serious consequences in the example of heart monitoring equipment shutting down or a process controller in a chemical plant shutting down a valve and creating an explosion. Y2K is not just a legacy problem and it is not just found in the back office systems of old companies.

While the Y2K problem may not be immediately apparent to unsuspecting bystanders, in time it could have ugly repercussions downstream for everybody. Even if the turn of the century doesn't go this badly, at a minimum, office and building automation systems will fail, elevators won't budge, buildings will be cold because the heat didn't come on and phone systems won't work.. Electronic time clocks, security systems, parking lot gates and bank vaults all freeze up. Fax machines go screwy. But, worst of all, the automatic coffee machine won't make coffee. However, it is the ripple effect that will have the most devastating impact on the unsuspecting and unprepared enterprise.

The Ripple Effect

Every electronic link is at risk. If one link fails, an entire supply chain is affected. This is the ripple effect of bad data. An old sailor's proverb says that: "In deep water, a tidal wave is only a few inches high."

The Y2K problem can be declared so pervasive that even if you may be Y2K ready, somebody else won't be. All of the confidence that you gain if your systems work but still are not that much good if everybody else's melt down. If the network goes down, business stops. Y2K will start a flurry of lost business leads, reactive activities and fire-drill decisions. The chain of liability is the result of the ripple of bad data that starts with the corruption of databases underlying industry.

In networks, data hops from one router to the next until it finally reaches its destination. If a router senses that a particular address stored in the route table has not been used for a specific amount of time, it will delete the entry from its database. This widely used process is called "aging". Routers that are not Y2K ready face the possibility of "aging" their entire database, by nearly one hundred years in one night. When the turn of the century comes, will all of the LAN passwords have expired or will the automatic purge of email messages older than a certain age have just caused the server to wipe itself clean?

We are now facing an inevitable socioeconomic disaster which will be felt around the world: the oil industry, the chemical industry, medical clinics. Would you think that thc Y2K problem is dangerous in the cockpit and air traffic control system? And, you may want to avoid being hospitalized – It just might save your life. Even if your enterprise can fix its Y2K problems, it surely won't get past the inevitability that state, local and federal information systems won't be ready.

Small companies and independent units of enterprises have a particular vulnerability to Y2K. And, even though they are the most endangered by Y2K, industry peers and trading partners could work harder together by "pulling them into the lifeboats." For the most part, the big corporations and enterprises are Y2K ready, but in the process they have consumed all of the resources it took to fix it, like programmers and money. The Gartner Group has identified about 3,000 Year 2000 vendors, of which only 200 have "de-bugging" tools for these same small companies, which run on everything but the type of programming code and languages found on the big mainframes such as COBOL.

The vulnerability to the Y2K problem will probably disrupt electronic commerce and internet web site operations. The Y2K bug will surface to scuttle and cripple parts of the internet. Although the internet is, in theory, self-healing, it could take a while and cost a lot of money. Call center is too visible a locus of failures, if you lose communication with your customers then you lose a competitive advantage because the worst signal you can send to your customer is a busy signal.

Tangible date crashes are already occurring, delivering early warning signals about how nasty and pervasive Y2K will become. These types of problems are already showing up and are not waiting for "the stroke of the millennial midnight." Everyday examples of things that can go wrong include staff scheduling, project and campaign management schedules, scheduled backups and financial documents and many others.

According to one expert, the optimal time for a mid-size corporation to begin Y2K readiness was 1995 or earlier. "October 1997 was the last point that a mid-size corporation could commence with repairs and have any hope of being finished by 2000." Electronic commerce in its current form has become a collection of disparate, proprietary and networked hodgepodge of systems.

If all of the worlds orphan systems fail or are shut down, chaos will ensue. If the experts are worried and having problems, what about the rest of us? That is, assuming that the elevators are working.

Several Simple Statistics

The Y2K Bug is expected to cripple at least 30% of the world's computers. One of every three people reading this paper will be affected significantly. Good luck, I hope that it is not you.

There are many consultancies and media hipsters that are fanning and fueling the Y2K fire. Finding believable and credible estimations of the real cost of Y2K among the flames is difficult to do. Y2K, in this sense, is a moving target. The Gartner Group, a telecommunications technology consultancy, has the biggest mouth on the block right now and according to them:

- Devices and systems built before 1996 have a 90 percent chance of experiencing some type of Y2K problem without intervention.

• On December 31, 1999, 30 percent of all externally-focused systems, such as call centers, and 50 percent of all other types of computing systems will have failed to achieved full Year 2000 compliance.

• Y2K will cost as much as $600 billion to $1 trillion with litigation costs.

• 90 percent of email systems won't work right with the Y2K Bug, resulting in misdirected messages or rejected messages.

• Nearly one-third of companies with fewer than 2,000 employees have not yet started to deal with the Year 2000 problem.

For a typical Fortune 500 company, 80 percent of business applications are date sensitive and are susceptible to some degree of failure. Current research data indicates that at least 15 percent of all software applications in the US will not be repaired in time. For most of you, it is not that the Y2K bug cannot be fixed, it is that there is a shortage of people that know how to fix it; experts in the "old languages" are getting paid as much as $1.00 per line of code (a good programmer can "scrub" several thousands lines per day).

In a survey of America's 128 largest corporations:

• One in five had not yet started a full-fledged remediation program.

• 37 percent reported computer failures occurring because of the Y2K problem.

• 65% percent of companies are reporting failures when they test for Y2K.

• 85% also admitted that they have underestimated the cost to fix the problem.

• 95 percent of companies expect the Y2K Bug to damage business flow.

In spite of estimates that as much as 75 percent of equipment problems will fall into the nuisance category, according to the Meta Group Consultancy, they expect that up to two-thirds of their clients will replace outright rather than fix their software and hardware for Y2K.

Without intricate planning and management the chances for success are low. History is on Y2K's side. In the past ten to fifteen years (in other words historically in the computing age), 15% of all large systems projects have been delivered, on average, 12 to 24 months late. About 25% have been canceled before completion. Why should the information technology track record be any different this time around, especially since the Y2K project is ten times bigger than any project any organization in history has ever done?

CIO Magazine found in a recent readers survey that nearly all business executives will not be flying on the Y2K date. What do you suspect they know?

The Economics of the Y2K Ripple

Y2K is no less than a global economy threat. If one link goes out, Y2K could spawn some unprecedented market disruption and the ripple effect could spill out into the foreign exchange markets and over to other markets, draining up to $10 billion in as little as five days. After all, the global currency market is utterly dependent on technology to transact. The Y2K success of banks will mean little if the global economy breaks down. In fact, success with Y2K will be meaningless if the economy fails.

Gartner Group claimed at one time that there is a 70 percent chance of a global downturn. Economists answer rising concerns that the odds of a severe global recession are 40 percent to 60 percent – a very serious threat to the US economy that will have a ripple effect and is bound to disrupt the global economy or vice-versa.

Downstream, Y2K recessionary trends and influences could last at least 12 months or more. The recessionary influence of Y2K is expected by some to be as severe as the 1973-1974 global recession caused by the OPEC crisis. Then, oil was the most precious resource, today it is information. If the supply of information is interrupted, many economic activities will be impaired, if not entirely stopped.

In general, the more an organization's systems affect daily cash flows and the greater the effort required to fix the Year 2000 Bug, the higher the risk that the organization will experience serious financial loss from the date change.

Based on respondents who have revised their Year 2000 estimates to include the desktop and networking software, it now looks as if the cost of bringing distributed computing into compliance will be at least half the cost of fixing mainframe systems. Even after completing a key inventory of fixing an enterprise's main computing systems, the major expense is emerging in the fixing of an enterprise's desktops and client/server systems – it is the chipmunks, not the elephant, where the greater Y2K problem lies.

It is likely that today's technology growth is being borrowed from spending at some time in the future, but certain industry segments and economic regions are more susceptible than others.

Segments at Risk

No company is immune from the Year 2000 problem, although, as with most illnesses, some populations are more at risk than others. But, are you really affected?

Some industries are more ready for the Year 2000 problem than others and some industries are more at risk than others. In the private sector, banking and healthcare are at the top of the list of most susceptible industries. But in particular, many are especially concerned about Y2K viability of banks and financial service enterprises. One industry most dependent on call centers is the financial services industry. Software and service providers are not that far behind banking and financial services when it comes to the Y2K problem.

At highest risk are those companies that have the highest dependence on operation systems to daily cash flows. Telecom and finance have close connections because of the required data flow. An application example includes the calculation of benefits for longtime employees and the execution of electronic transactions. Reservation systems that accept bookings one year in advance and business planning systems that project more than one year in advance such as those found in hotel operations may accept corrupted records and begin billing guests for 29-year-long telephone calls or simply reject the transaction and fail to bill the guest at all.

There are regions also with warnings: Southeast Asia, Latin America and the Middle East. These are regions where large companies or entire governments will not be ready. In many countries, absolutely nothing at all is being done about Y2K. If your enterprise does business or runs call centers globally, you will find that the Year 2000 problem is much more serious and acute, particularly in Latin America. Many of these countries use older systems and therefore have more code to fix. The fact is that the US accounts for only 16 percent of the world's computer code. There's less oversight of the Year 2000 problem outside the United States and certain cultures have a reluctance to admit something might be wrong. For a variety of reasons, companies overseas - particularly those in Europe – are struggling twice as hard and yet are further behind.

In Europe, the Year 2000 problem is competing with other information technology resources and high priorities - particularly those related to the conversion to the new Euro-Dollar currency. Experts are estimating that the Euro-Dollar conversion is four to five times more complicated than the Year 2000 problem. The net effect is that the Year 2000 problem is taking a backseat to Euro-Dollar currency conversion and as a result, European call center operations are particularly vulnerable.

The Government, Y2K's Best Enemy

This brings up the point that - and it is no different in the private sector - if the fundamental systems of an enterprise are in disarray, the downstream systems found in the servicing of customers, particularly the call center, are certainly affected as well. The biggest industry segment to be affected is federal and state and local government entities. Even if your enterprise can fix its Y2K problems, it surely won't get past the inevitability that state, local and federal information systems won't be ready.

The problem is serious enough that the US House and Senate, the Federal Reserve Board, the Securities and Exchange Commission, the Comptroller of the Currency and the White House, among other federal agencies, have weighed in on the issue - particularly ironic since the federal government (especially the IRS and the Social Security Administration, which run some of the largest call center operations in the world) is the least prepared for Year 2000. Experts have been giving D's and F's to government agencies for their progress.

Government regulatory involvement may make it impossible to use any better solutions that may evolve as conversions advance. Government agencies, in the ultimate regulatory irony, are themselves the most likely to fail on Y2K.

To put some perspective on the enormity of the problem, the federal government's Office of Management and Budget estimates that the price tag to avert widespread government computer crashes from the Year 2000 Bug has grown from $2.3 billion in February, 1997 to $2.8 billion in May, 1997 to $3.8 billion in August, 1997 and $3.9 billion in January, 1998.

The cost of US government fixes will exceed $3.9 billion. A new estimate was for $4.7 billion for 24 government major agencies. For the Canadian audience that are reading this paper, the Canadian Government will spend about $1 billion to fix their Y2K problem.

Few people would argue that the ultimate and most mission(critical "customer contact center" is an air traffic control tower. The FAA (Federal Aviation Administration) delivers very good customer service considering that they use in a majority of their Air Route Traffic Control Centers thirty ancient IBM 3083 computers manufactured in the early 1980s. They take radar information and convert it into visual display data reporting an aircraft's location, identity, altitude, speed and destination. The 3083 computer has, at most, all of the power of a small file server today. The situation at the FAA is so bad that they do not plan to be completed until mid 2009 and this author could only find one lone Federal Aviation Administration Y2K official would admit that he would be flying on December 31, 1999.

The IRS was one of the first to "be stung by the fallout" of Y2K Bug. The Internal Revenue Service is expected to spend over $900 million to scour 62 million lines of computer code, second only to the Defense Department as the costliest Year 2000 problem fix in government. The IRS has already had Y2K experience in which 1,000 innocent taxpayers received an incorrect notice of late tax payments during tests.

I'm not sure I would trust a CIA that missed not one, but several nuclear bombs being detonated in the far east, but a recent Reuters wire story reported that the CIA employees have been advised to pay their bills early in December 1999, to have cash on hand and extra blankets and stores for an extended blackout.

Y2K will probably hit close to home as well. Fire and police dispatch systems (another truly mission-critical call center) are exposed. The vast majority of police and fire equipment are not Year 2000 ready. On January 1, 1997, the Y2K bug "arrested" a police department computer responsible for criminal records, driver's licenses and vehicle registration and wouldn't let police set court dates further than the Year 2000.

When it comes to the federal, state or local government providing a solution, not only can they not fix theirs, but I, like most of the American public, have more confidence in Bill Gates than in Bill Clinton.

Finance

No one is putting the telecommunications industry under the microscope like banks are being examined, which may lend a certain vain twist on our cultural psyche. Investors, shareholders and stockholders are taking the first steps to recognizing and sorting out the companies that are ready to survive the Y2K bug and those that are not. As a result, there is an increasing pressure on companies to disclose details of their Y2K efforts.

With a Y2K expiration date, credit cards are being rejected by retailer's terminals and POS networks all over the world. This will cause a ripple effect of record numbers of cards being rejected and spit out, forcing retailers and merchants to key in and call in for authorization. With credit cards, there is a small segment of merchants with terminals that are causing the Y2K problem, demonstrating that a little Y2K problem can go a long way. The consequences that we are witnessing in the credit card industry are not business killing, but it is now definitely obvious that that we are having Y2K problems.

Y2K problems may be as simple as customers accruing 18 percent compound interest on credit cards over 99 years or, like the one at a Fortune 500 financial services company in the midwest, in which the company's consumer loan systems encountered the "00" date and sent 200 customer bills for 96 year's worth of interest. A Wall Street brokerage firm made a Y2K testing mistake and accidentally deposited $19 million in each of its client's accounts.

The banking industry has 275 major processing services and 12 major software vendors. The Bug will leave banks unable to post check payments (wrong date) and accurately calculate interest and amortization payments. ATMs will refuse to work. The bank phone systems, ACD and security systems will all be stalled, not to mention elevators, fax machines and worst of all the coffee maker.

Alarmists are particularly concerned that, as word gets out about how unprepared banks and other financial institutions are, the Year 2000 Bug could conceivably cause a run on banks. In a telling vote of confidence in their own systems, of 1,100 IT banking professionals surveyed, 38 percent said that they may withdraw and make liquid their personal assets from banks and investment companies shortly before Year 2000. For some time, the Securities Industry Association was seriously considering making December 31, 1999 a trading holiday so that enterprises had an extra day to complete year end processing before Y2K rollover, although that would require an unlikely executive order from the President.

The problem may be so acute in banking and finance that even federal and state governments are getting into the action, threatening to severely reprimand banks that failed to meet a September 30, 1997 deadline for detailed plans for preventing Year 2000 snafu's. In Georgia, the Federal Reserve has issued cease and desist orders to three banks for falling behind in their fix-it efforts. 105 more banks have been warned. As many as 700 banks are expected to close their doors if the check clearing software breaks down to the Y2K Bug.

If you are a publicly traded company, the public embarrassment and hit to the stock price that a widespread Y2K problem could cause may be horrendous.

The FDIC has insured about $2.7 trillion held in about 11,000 US banks and, although the overwhelming majority of banks are scheduled to have completed their Y2K readiness, it is the estimated 2,000 small and mid-size banks that can expect catastrophic difficulty. In a twisted bit of irony, the very same agency that wants to close several banks for not being Y2K ready is itself behind its Y2K readiness schedule by more than eight months.

There are even insurance plans available for the Year 2000 challenge, but the premiums are $65 million to $80 million for each $100 million of coverage - an indication of how much faith the insurance industry has in Year 2000 preparedness.

Vast and huge segments of the information intensive financial service market may be struck silent by Y2K bug. The danger with the Y2K bug is not the fatal silence that it could put on the financial and banking service sectors, but that even a minor wound could prove deadly should you be in a hospital on New Year's Y2K.

Medical

In the mid 1980s, a software bug caused overdoses in a Therac-25 radiology machine that accidentally killed several people. This was caused by a bug very similar to Y2K bug. The healthcare industry is very susceptible to Y2K, from blood-testing devices, diagnostic machines to pace makers - the ultimate embedded chip – are all precariously dependent on time/date functions.

One expert told The London Times that a 10 percent failure rate in compliance in the healthcare industry would result in 600 to 1,500 deaths. The Gartner Group has recently suggested that more than 80 percent of hospitals world wide have failed to address Year 2000 problem.

And, while life support systems and medical diagnostic equipment are being diligently scrutinized and tested, it is the patient billing and record systems (the very same systems that deliver information to healthcare provider call centers) that are woefully behind in preparedness for the Year 2000 problem. Gartner Group claims that nearly seven out of ten healthcare groups face the risk of systems corruption and collapse from insufficient response.

There is a general feeling that the healthcare industry has been particularly slow in picking up the responsibility to determine the impact of the Y2K on mission-critical life support technologies such as ventilators – another life support system. An acceleration in the merger and acquisition of hospitals and health care services exacerbates and slows the remioediation of the date-sensitive bugs in the computer systems tracking everything about a patient from records to accounting.

There is no more mission-critical customer service than that found in a hospital. What the industry needs is to communicate and produce brand new bug free equipment. As many as one-half of a hospital's vendors are being silent about whether, or how, they are fixing their systems – a foreboding indicator. This is truly an industry where vendors must communicate their Y2K readiness.

Utilities

Power companies have thousands of embedded chips in their equipment, and the fact is that most of them will not be tested before Y2K. There is a real possibility that we will have blackouts throughout North America.

By July 1999, all nuclear plants must be ready or they must shut down for safety reasons. The Nuclear Regulatory Commission anticipates having to shut down more than ten percent of the nation's nuclear plants because their systems cannot handle the Y2K rollover. NRC has ordered 104 nuclear plants to file compliance reports. There are nearly 8,000 companies distributing or generating power in the US and the frequency of embedded devices is high. Is there any irony that the first major buyer of a Y2K ready plug-in chip that replaces the clock chip in millions of places is the US Nuclear Regulatory Commission.

The Phonet

"Year 2000 Bug threatens phone service", shouts a USA Today headline. Failures will range from billing problems to a complete lack of phone service. Telephone networks are computerized at every level, from the transmission of calls to billing and ordering of supplies. There is a fifty percent to sixty percent chance that each major telecommunications carrier will suffer the failure of at least one mission-critical system. Like power, we can survive a short while inconvenienced by the lack of dial-tone.

Phone systems are particularly vulnerable to Y2K from "bad BIOS date" problems which can interfere with peripheral reporting and management systems that basically produce garbage SMDR records. In telecom systems, the problem is not exactly the real-time switching of traffic but the corruption of data used in time stamping, call-detail records and billing. In telecommunications for example, voice mail systems can exchange messages, but not the accompanying set of features. ANI, telephone numbers and query identifiers can not be shared among disparate systems.

So, even if all of your own systems and applications survive the rollover, there is still the fact that they probably won't work with anybody else's. If there is any industry steeped in old code, it is the telecommunications industry. There is an underestimated market of "old" non-Y2K-ready telecommunications systems out there that cannot jump the Y2K hurdle. Up to 60 percent of network carriers are expected to go dark as parts of their networks wrestle with odd date/time anomalies. There is a darn good chance that everybody's telephone bills will get so hopelessly messed up and mangled that none of us would have to pay. They essentially become audit proof.

Telecommunication is a market segment built entirely on high volume transactions that are measured in detail by their date/time and duration. Date/time is used for routing, costing, billing and rebilling. The phone system and its peripherals are all programmed to have events based on dates, time of day and day of week. While it is safe to say that no telecom manufacturer is unaware of Y2K, many have yet to express definitive guidelines on how they plan to service Y2K problems.

None of the three major network systems vendors, IBM, Microsoft or Novell are guaranteeing their products against Y2K problems. At last count, the internet itself needed software upgrades to about 300 routers and switches. Y2K is buried in routers, hubs and even firewalls themselves of both internal and external corporate networks.

Which particular computer telephony systems can be affected? All of them; however, voice-mail systems, auto-attendant, off-board reporting and custom IVR systems are the most likely candidates. The telephony and call center industry is unique in that there are still many seven - to twelve-year old legacy voice-mail and IVR systems out there and operational - in fact, some still exist that date back to as far as the mid to late 70's. In general, the core telecommunications equipment market segment, including ACDs, PBXs and key systems, is large because systems are expensive and they generally have longer lifespans.

As a serious note on the Year 2000 problem, there have been a number of articles published in the telecommunications and computing press over the past several months insinuating and implying that the Year 2000 problem is "vendor-created." This is bull. Many call center equipment vendors will acknowledge that if they could have artificially created a demand for their products and services, they would have certainly chosen a less risky area.

There isn't much back up available for telephone service and a prolonged outage would be a "show-stopper" for every enterprise, particularly those that believe they will transact more business over the phonet. Plenty of enterprises are in for a nasty surprise on Y2K day when they can not receive their customer calls.

The Call Center's Own Y2K

Somewhere in every reader's business food chain, there is somebody taking care of the customer.

Civilization won't collapse but customers may be annoyed. At a minimum, average call handling times will skyrocket as agents must take time to explain to callers why account information or order processing information isn't available. The damage will be a loss of confidence in your call center as the first impression customers have of you. Customer service agents may be doing a lot of explaining. If, come the Year 2000, they cannot do their jobs, they are the ones that are going to take the heat and it's the call center manager's job to ensure that the image desired by their enterprise is carried out.

While most managers responsible for call center operations are aware of the Year 2000 problem, a recent survey of 1,000 senior executives found that 13 percent were not even aware of the Year 2000 problem, 16 percent were still deciding on a plan, 30 percent had a plan but had not begun implementation yet and only 40 percent had begun fixing it. Of the 60,000 to 70,000 call centers with some four million agent seats and millions of PBXs in the US, the potential for disaster is significant.

Unlike the computing industry, which is seemingly in a perpetual state of upgrade, call center managers tend not to upgrade technology unless it is broken or hopelessly out of date, with only 50 percent of call centers regularly upgrading their equipment, according to Dataquest. Also according to Dataquest, most of the installed base of ACDs is at least two software releases behind.

According to the Gartner Group, practically every function on a PBX or an associated system is ultimately controlled by the date, and the fact is that many call center managers really don't know how their existing systems will respond on January 1, 2000. Unfortunately, an informal sampling has shown that many call centers don't understand the implications the millennium date and time change will have on their voice operations.

Unless prepared for, havoc will occur the first time date-related digits beyond the year 1999 are entered into call records, workforce-management software, call-routing software, order-processing applications or other databases. This is a particularly acute issue with reporting-application software resident in off-board report generators for ACDs. Date incompatibilities could cause problems in the sorting and processing of reports and messages in folders or discussion databases, auto-aging and auto-deletion of reports and routing schedules, calendar entries, automatic rejection or deletion of messages and folders, the logging of messages and rules-based workflow applications. The Year 2000 problem could mean that critical messages or reports may never be seen by their intended recipients.

Most major call centers have extensive highly-articulated systems and applications, making them particularly susceptible to "one-off" software problems. They are difficult to test while in service and are very expensive to bring down. Anything that manipulates SMDR data may fail. Some call center applications will try to perform statistical or time-stamped reporting using strange date spans. ACDs are particularly vulnerable because of conflicts with scheduling and routing, management terminals, reporting and analysis systems and any other application that requires a password. Certain astute call center managers are acutely aware of this and in many organizations they are extremely concerned.

If there ever was a technical issue that call center managers were avoiding ("it's just a data problem"), this is it. The point was succinctly made that, while ACD vendors have gone to extraordinary lengths to insure that the ACD does not fail at the turn of the century, it is failure with all of the other peripheral devices, systems and applications that will create havoc. Your ACD may continue to take calls, but if your agents can't get to the call center because parking lot gates and elevators won't function or they can't process calls because their desktop PCs have crashed this is not a data problem, it is your problem.

Tests have shown that simply being able to display January 1, 2000 on reports, supervisor consoles and agent workstations isn't enough. Although the ACD may handle the Year 2000 challenge well, it is important to understand how third-party peripheral products such as voice messaging, call accounting, ACD and MIS reporting applications, among others, will not handle the change.

Many call centers are neglecting to check for Year 2000 preparedness of desktop applications, many of which have been tweaked and adjusted by the agents and their supervisors. From this perspective, these desktop systems may be more critical to the business than the enterprise mainframe. If agents cannot get access to the information that they need to do their jobs, you had better believe that there is going to be a problem and that is why call center managers should be concerned. Calculate how much money your enterprise loses if all of your screens go down and you have to pay everybody just to be there.

Prepare paper-based scripts now. Prepare an agent script on paper telling callers what has happened in anticipation of agent desktops not functioning properly. Most of the major carriers and the PSTN will be prepared for the Year 2000 Bug, so expect calls to be delivered to the call center. It is when they get there that difficulties can be expected.

At a minimum, expect some sort of whack on the call center head from the Big Bug. Speaking directly to the customer service and call center market, Dataquest, a San Jose-based research group (a division of Gartner Group) has estimated that as much as 25 percent of the 1994 installed base of PBX-based ACD systems may need to be upgraded to handle the Year 2000 problem.

Telecommunication applications are the result of multi-vendor integrations. A small PBX/ACD may have five or six distinct components such as a mother board from one vendor, a fax board from another, a voice board from yet another, software and operating systems from others and so on. At the enterprise level, call centers have the ironic misfortune to be the crossroads of so many, often disparate and different computer and processor-based technologies, PBXs, ACDs, dialers, reporting systems, IVR, voice messaging, network routers and databases are examples. The greatest risk is not just the corruption of transactional data. Even if these systems don't fail, your call center won't work if your agents can't get into the building.

Retrofitting and recompiling all of the heritage applications in an enterprise is proving to be tremendously time consuming, a process riddled with errors and very expensive, incomplete documentation. An effect of the Year 2000 problem that few in the customer service and call center industry will admit, is that some call centers are using Year 2000 Bug as reason to replace systems outright. Replacing systems is an absolutely certain way to eliminate any problems from the Year 2000 Bug.

The Vendors and Y2K

The decision to communicate or not about Y2K is being made by marketing departments, many of whom are unintentionally completely clueless. A by-product of the information age has been the co-dependance on software application builders missing and slipping on delivery dates (Windows 98?).

As the Y2K lurches closer, some enterprises are becoming alarmed at the number of vendors that are sweeping the Y2K problem under the rug.

Before IS comes knocking on the call center door, managers must be proactive and must assume that putting off the long-overdue confrontation with the Y2K Bug is no longer an option.

A serious note on the Year 2000 problem is the observation that there have been a number of articles published in the telecommunications and computing press in the past several months insinuating and implying that the Year 2000 problem is "vendor-created." This is not true. Most vendors have spent a tremendous amount of money - money preferably spent elsewhere - making sure that their products and systems are ready to meet the Year 2000 challenge. Many call center equipment vendors will acknowledge that if they could have artificially created demand for their products and services, they would have certainly chosen a less risky area.

The telephony and call center world is taking the consequences of the Year 2000 problem less seriously than those in the data world, although awareness of the Year 2000 problem is now certainly growing in the voice industry at a much quicker pace. In recent months, Year 2000 compliance has increasingly started to find its way into most requests for proposals (RFPs).

One of the tests for the Year 2000 Bug is to specifically ask your vendor if your PBX/ACD and related systems are programmed to recognize that February, 2000 will have 29 days (Year 2000 is a leap year whereas the year 1900 was not). If your systems recognize the date February 29, 2000, then it is almost certain that your call center systems have the ability to make the Y2K change.

Just asking vendors for written guarantees that their ACD/PBX telephone systems are Year 2000 compliant may not be enough to prepare for Y2K. Only 21% of IT professionals say that they will offer warranties or guarantees to be Y2K ready. Most of the established ACD equipment vendors have thoroughly prepared their systems for the Y2K Bug and if their systems have problems, odds are they already know. Many vendors have bundled Y2K solutions and fixes into the latest release of their products, assuming that companies will want the latest functions and features. What they may have failed to consider was the time it would take to move to a new release. Unbundled Y2K fixes may be the best offer.

In several cases, vendors are coming to grips with how much time and expense it is costing them to make their software Y2K ready. It is unrealistic to expect vendors to agree to fix their systems for free in exchange for liability limits, since most third party software is already heavily customized and requires substantial resources to repair.

Be aggressive with your vendors and start asking questions now. Do not rely solely on the word of your vendors, but don't assume that they are not afraid of Y2K either and are doing their smartest best to minimize the questions. For those forward thinking companies that attempt to stay up with technology and prepare their customer service operations as Y2K nears. perhaps the lesson to lbe earn here for those that don't, is to pick better business partners.

Find out if your primary ACD equipment vendor will fix equipment it supports from other third party vendors. Determine what your vendor will do for free and what needs to be upgraded; get it in writing. Test upgraded systems well before the Year 2000 deadline and if in doubt, do exactly what is recommended by the vendor (you may end up with a few new functions and features) or spend the money to retain competent professionals or consultants to do testing.

Legal

The legal ramifications are staggering and, while your enterprise believes it is prepared for the Year 2000 problem, there will be a certain amount of the domino effect. What a big company considers unimportant may be your company's lifeblood.

Many vendors have spent a tremendous amount of money - money preferably spent elsewhere - to make sure that their products and systems are ready to meet the Year 2000 challenge.

Government systems that they interact with will be besieged by lawsuits from retired or ill workers as their compensation, insurance and pension payments become misplaced and mis-processed. Blanket protection is not fair to those companies that have already invested plenty of resources to minimize the impact of Y2K.

Legal experts are predicting that Year 2000-related litigation could go beyond the $1 trillion mark (that's twelve zeros). Y2K has turned into a legislative codependency program for politicians. Legislators in California, Nevada, Utah and several other states are pushing legislation to protect their states' respective high-tech industries from being devastated by lawsuits and to restrict punitive damages awarded in Year 2000-related lawsuits. A recent bill was introduced in California to limit lawsuits resulting from Y2K failure. Lawyers and litigators are lining up in anticipation of many problems from Year 2000. The supposition is that, with two years to prepare for Year 2000, vendors that don't conduct due diligence to repair their systems should be sued.

Many companies are sending out letters that are, for the most part informational, but have embedded in at least one paragraph some admonition like: "In our view, failure on the part of any company to prepare for and resolve Year 2000 failures constitutes negligence and, as with all acts of negligence, exposes the negligent party to claims for damages. We will not hesitate to aggressively prosecute Year 2000 system failure-related claims."

When requesting additional copies of system software for testing purposes or when testers need the vendor to provide a "software key" or access code to test the software for dates beyond the expiration of the software license, vendors may attempt to charge additional licensing fees. Most of these disputes are immediately and amicably resolved with vendors; however, if there is any hesitation by your system vendor to facilitate testing vigorously, remind him that the legality of doing so is questionable.

The problem with Y2K, as Michael Schrage pointed out in a recent column, is that the problem isn't with ignorance at this point, it is with arrogance. "We know what the problem will be: they just do not believe that it is going to happen to them."

One unexpected challenge that many call centers are finding when performing Year 2000 testing is that they may have to test the systems on a separate or bigger processor than the current software licenses may allow. A few call centers that have started testing their Year 2000 software fixes have run into a snag: software licensing disputes with vendors. Be aware that with "move-ahead" date testing, you may have software licenses that expire before the testing date and the testing won't work. Software vendors in particular are increasingly being warned that they must provide free upgrades to the older versions of their software.

Use lawyers now while they are cheap. A lawyer can make it very, very clear in a letter to each of your application and platform vendors that your call center will be making business decisions based on the information that the vendor gives you about its system's Year 2000 readiness and that therefore the vendor will he held liable if the information it provides you is incorrect. Use some legal weight to get some honesty and assume that your vendors and your systems are guilty until proven innocent.

An unfortunate effect of Y2K are the movements afoot to encourage corporate employees to blow the whistle or to publicly expose Y2K slackers and laggards. There are plenty of tipsters anonymously blowing the whistle on companies that are ignoring their Y2K responsibilities and problems. The bottom line is that it is the responsibility of knowledgeable managers to notify the enterprise's legal counsel when the enterprise may be exposed to a potentially serious problem.

Conclusion

Never before in history has software's vital importance to business been so visible and vulnerable. The Y2K problem definitely qualifies as the most expensive industrial accident of modern times. Or, Y2K could well turn out to be possibly the biggest scam ever to be perpetrated on the country by consultants.

Readiness for Y2K is a binary issue, not one with a spectrum of choices. And, although planes will not fall from the sky, nuclear reactors will not melt down (as many people would be surprised to find out, most nuclear reactors are not even operated by computers) and financial institutions will not freeze. While signs of the Year 2000 problem may not be cataclysmic, such as airplanes falling out of the sky or nuclear reactors melting down, most people will certainly see things like account balances at the video store showing late charges, auto dealerships sending notices that their cars are years out of warranty, customers being billed for things that have already been paid for and, for others, things will likely just simply degrade. In spite of all of these doom and gloom headlines, public awareness is increasing slowly.

Blissful optimism is ignorant at this point as well. It is stupidly naive to expect that all systems will roll over and die gracefully on January 1, 2000. Many will continue to run, only delivering bad dates and times that will rapidly corrupt the enterprise databases. A company may still be able to carry out its core business functions if a non-critical system fails, but the enterprise can be severely impacted if it is blocked all together. A mission-critical system for a hotel, for example, may shut down the elevators, a non-critical systems failure may prevent guests from getting their wake up calls or misstate management reports. The most insidious effect of the Y2K Bug will be bad data creep.

The consuming public, your customers, will want answers if the Y2K leads to failures and service interruptions. They sense the frustration that is mounting among technology professionals fueled by the academics, to pressure vendors to meet deadlines and deliver on Y2K readiness promises. Scary reports are written by people who obviously do not have a familiarity with how computer systems work. This makes them easy prey for hype-y scare tactics of the media, computer consultants and the entire technology food chain.

While failure is not exactly catastrophically imminent, there is now an opportunity to emerge from the effects of the Y2K bug with a major shift in the use of telephony and software services. In fact, the crash of Y2K may be a good thing. There is a big chunk of hardware, software and the applications that they are built with that does deserve to become extinct. It just may be more fortunate than not that the systems most affected by Y2K are the oldest and are the most deserving of this proverbial "last nail in the coffin." It is quite possible that the savings of doing it the old way might have been greater than the cost of repairing the damage from Y2K now.

A majority of people simply believe that we should stop being alarmists about Y2K and that companies in a free market can handle whatever complications from Y2K arise and that the market will rebound and readjust from any level of calamity. Certain segments of the economy will actually be profitable and robust.

As long as whole industries do not fail, then those suppliers of goods, services and technology will pick up the slack and eventually provide enough supply to satisfy demand. What can't be predicted or avoided is the downstream effect of the Year 2000 problem - the Fortune 1,000 will stumble significantly if the Fortune 1,000,000 customers can't buy their products.

Whether you are a customer, executive or a vendor, you are already entangled in the Y2K net, hype or not. The customer, the manager and the vendor each have a different take on Y2K . Understand each's perspective and you will have a greater chance of bringing your enterprise through - ripple free.

To give you an idea about what the consuming public is thinking, consider that a recent study said that the public's confidence was greater in high-tech companies than in lawyers, government and insurance companies. Bill Gates has higher public opinion scores than Bill Clinton. We have all heard now from plenty of experts the prediction that a substantial number of enterprises will experience preventable costs, revenue losses, and in worst case scenarios, complete business process failures.

The only way to effectively communicate the seriousness of the Y2K issue to executive management is, unfortunately, scare the shit out of them. At least you've got a few months to prepare for this one. Be proactive with your vendors and use a lawyer while they are cheap. Communicating with customers about Y2K readiness is vitally important as well. Johnson and Johnson's 1982 handling of the poisoned Tylenol demonstrated remarkable results when the crisis called for candor. Deception and special legislated protections are not the right long term strategy here, either.

For vendors: a famous Russian said, "Trust, but verify."

In the real world, there are those companies that are handling the Y2K conversion swiftly and efficiently. Fast, clean and crisp technology choices are being made everyday. Let those that have expired from Y2K retool. For the most part, many anticipate the Y2K will just turn out to be the biggest party ever when people realize that we are still here with some minor bumps. For others, some merciless decisions will have to be made in deciding which systems and users will survive and which ones will not.

Companies must plan to fail with their Year 2000 work and begin to make contingency plans in anticipation of some part of their call center operations failing on January 1, 2000. So you must have contingency plans, the same kind of business recovery strategies you've prepared for other disasters. Even if you fix yours, the chances are that somebody didn't fix theirs.

Customer service technologies, specifically those found in call centers will benefit the most from the Y2K problem. If your technology works, be prepared to deal with the flood of calls from customers with questions about the parts and products of your enterprise that did not survive Y2K unscathed. The fall back interface for society will be the telephone. Are you Y2K ready to take all of those calls?

Any recessionary trends in the future, particularly global, will see a surge in customer service activity, not a shrinkage of demand. In recessionary Asia, the computer-telephone and other service engagement-related technologies are growing quite significantly. In particular, look for Y2K opportunity segments and growth in call center and telemarketing service bureaus, PBX-based ACDs and Microsoft NT-based applications.

The real flashback from Year 2000 will come months, maybe years after 2000. Be careful about winning the battle but losing the war. The Year 2000 Bug won't just be a big problem for the first two or three months of 2000 and then simply go away. The ripple effect of bad data may also make many feel as though they have conquered the millennium-date-field conversion problem, but will find that they lost their technology edge and then, eventually their jobs.

For customer service and the call center technology and applications, Y2K will likely prove to be more of a nuisance than a neutron bomb, creating minor inconveniences for customers and employees downstream. As catastrophically crippling as the as the Y2K Bug may be, the trick will be in making its impact a non-event to your customers.